AF774444

Springer Proceedings in Mathematics & Statistics

Volume 536

This book series features volumes composed of selected contributions from workshops and conferences in all areas of current research in mathematics and statistics, including data science, operations research and optimization. In addition to an overall evaluation of the interest, scientific quality, and timeliness of each proposal at the hands of the publisher, individual contributions are all refereed to the high quality standards of leading journals in the field. Thus, this series provides the research community with well-edited, authoritative reports on developments in the most exciting areas of mathematical and statistical research today.

Cheikh Thiecoumba Gueye · Papa Ngom · Idy Diop
Editors

Mathematics, Computer Science, Cybersecurity and Artificial Intelligence

6th Scientific Days of Doctoral School of Mathematics and Computer Sciences (S2DSMCS), Dakar, Senegal, December 17–19, 2024

Springer

Editors
Cheikh Thiecoumba Gueye
Département of Mathématiques et Informatique
Cheikh Anta Diop University
Dakar, Senegal

Papa Ngom
Département of Mathématiques et Informatique
Cheikh Anta Diop University
Dakar, Senegal

Idy Diop
Ecole Supérieure Polytechnique
Cheikh Anta Diop University
Dakar, Senegal

ISSN 2194-1009 ISSN 2194-1017 (electronic)
Springer Proceedings in Mathematics & Statistics
ISBN 978-3-032-17665-3 ISBN 978-3-032-17666-0 (eBook)
https://doi.org/10.1007/978-3-032-17666-0

Mathematics Subject Classification: 11D61, 17A35, 94B27, 92D30, 94A60, 00B25, 65M60, 68T01, 60H05, 68M10, 68U99

© The Editor(s) (if applicable) and The Author(s), under exclusive license to Springer Nature Switzerland AG 2026

This work is subject to copyright. All rights are solely and exclusively licensed by the Publisher, whether the whole or part of the material is concerned, specifically the rights of translation, reprinting, reuse of illustrations, recitation, broadcasting, reproduction on microfilms or in any other physical way, and transmission or information storage and retrieval, electronic adaptation, computer software, or by similar or dissimilar methodology now known or hereafter developed.
The use of general descriptive names, registered names, trademarks, service marks, etc. in this publication does not imply, even in the absence of a specific statement, that such names are exempt from the relevant protective laws and regulations and therefore free for general use.
The publisher, the authors and the editors are safe to assume that the advice and information in this book are believed to be true and accurate at the date of publication. Neither the publisher nor the authors or the editors give a warranty, expressed or implied, with respect to the material contained herein or for any errors or omissions that may have been made. The publisher remains neutral with regard to jurisdictional claims in published maps and institutional affiliations.

This Springer imprint is published by the registered company Springer Nature Switzerland AG
The registered company address is: Gewerbestrasse 11, 6330 Cham, Switzerland

If disposing of this product, please recycle the paper.

Committee

Program Committee

Takoua Abdellatif, Polytechnic School of Tunisia, Tunisia
Lamine Ba, Ecole Superieure Polytechnique, Universite Cheikh Anta Diop, Dakar, Senegal
Mandicou Ba, Ecole Superieure Polytechnique, Universite Cheikh Anta Diop, Dakar, Senegal
Pierre Louis Cayrel, Universite jean monnet, Saint Etienne, France
Doudou Dione, Universite Cheikh Anta Diop, Dakar, Senegal
Amadou Abdoul Diop, Universite Gaston Berger, Saint Louis, Senegal
Idy Diop, Ecole Superieure Polytechnique, Universite Cheikh Anta Diop, Dakar, Senegal
Boudi Nadia, Mohammed V University, Morocco
Daouda Diouf Universite Gaston Berger, Saint Louis, Senegal.
Ibra Dioum Ecole Superieure Polytechnique, Universite Cheikh Anta Diop, Dakar, Senegal
Omar Kihel, Brock university, Ontario, Canada
Juan A. Lopez Ramos, Almaria University, Spain
Dolores Martin, Malaga University, Spain
Edgar Martinez Moro, Universidad de Valladolid, Spain
Salma Mesmoudi, Université Paris1 panthéon-Sorbonne, France
Souidi El Mamoun, Mohammed V University, Morocco
Ousmane Sadio Ecole Superieure Polytechnique, Universite Cheikh Anta Diop, Dakar, Senegal
Abdellatif Rochdi, University of Hassan II of Casablanca, Morocco
Abderrahmane Ntaj, Caen University, France
Papa Ngom, Universite Cheikh Anta Diop, Dakar, Senegal
Cheikh Thiecoumba Gueye, Universite Cheikh Anta Diop, Dakar, Senegal
Robin Quillivic, Institut des Systemes Complex (ISC PIF), Paris, France
Leila Mesmoudi, Universite Cheikh Anta Diop, Dakar, Senegal

Boudi Nadia, Mohammed V University, Morocco
Babacar Mbaye Ndiaye, Université Cheikh Anta Diop de Dakar, Senegal
Abdenacer Makhlouf, Université de Haute Alsace, France
Candido Martin Gonzalez, University of Malaga, Spain
Abdel Malek Azizi, Mohammed I University, Morocco
Ibrahima Fall, Ecole Superieure Polytechnique, Université Cheikh Anta Diop de Dakar, Senegal
Mountaga Lam, Université Cheikh Anta Diop de Dakar, Senegal
Diego Napp, Almaria University, Spain
Amadou Tall, Université Cheikh Anta Diop de Dakar, Senegal
Ioannis Chalkiadakis, institut des sciences complexes Paris Île De France (ISCPIF), France
Charles Tijus, université Paris 8, France
Paul Bourgine Complex Systems Digital Campus, France
Waldo Cancino Sao Paulo University, Brasil

Organization Committee

Idy Diop, Cheikh Anta Diop University, Senegal
Ibra Diuom, Cheikh Anta Diop University, Senegal
James Tamgno Kouawa, Multinational Higher School of Telecommunications, Senegal
Leila Mesmoudi , Cheikh Anta Diop University, Senegal
Papa Ngom, Cheikh Anta Diop University, Senegal
Ousmane Ndiaye, Cheikh Anta Diop University, Senegal
Demba Sow, Cheikh Anta Diop University, Senegal
Alassane Diouf, Cheikh Anta Diop University, Senegal

Preface

The 6th Scientific Days of School Doctoral Mathematics and Computer Science took place on December 17–19, 2024, in Cheikh Anta Diop University of Dakar.

They were organised by the School Doctoral Mathematics and Computer Science of Cheikh Anta Diop University of Dakar.

The 6th edition received around 34 submissions, and all were reviewed by the programme committee. Each paper was assigned at least to two reviewers. After highly interactive discussions, a careful deliberation, the programme committee selected 13 for publication.

The authors of accepted paper were given a week to prepare final version for these proceedings.

We would like to note that these scientific days offer a prestigious and conducive forum for researchers in the fields of Mathematics, Computer Science and Telecommunications to share their most recent findings, and engage with internationally renowned colleagues and experts, and in so doing contribute to the advancement of science.

The purpose of this scientific event is to stimulate research values, explore emerging challenges in science and bring to the fore the concrete applications of mathematics and computer science on modern society.

We are deeply grateful to the programme committee for their hard work, enthusiasm and conscientious efforts to ensure that each paper received a thorough and fair review. We also would like to thank Laila Mesmoudi for her useful help, as well as sponsors of the event: CEA-MITIC, CEA-AGIR and DCSSI at Presidency of the republic of Senegal.

Last but not least, we give thanks to all those who contributed to this event.

Dakar, Senegal
November 2025

Cheikh Thiecoumba Gueye
Papa Ngom
Idy Diop

Contents

Absolute-Valued Algebras with Left-Unit Satisfying $(x^2, x^2, x^r) = 0$ Revised

Pape Aliou Dieng, Alassane Diouf, and Abdellatif Rochdi

Abstract Let $\mathcal{A}$ be an absolute-valued algebra with left-unit. For such an algebra, the two identities $(x^2, x^2, x) = 0$, $(x^2, x^2, x^2) = 0$ are equivalent and give, in dimension 8, four algebras. This allows to have a non-redundant list of all absolute-valued algebras with left unit satisfying $(x^2, x^2, x^r) = 0$.

Keywords Absolute valued algebras · Left-unit · Linearization of identities

Mathematics Subject Classification 17A30 · 17A35 · 17A36 · 17A80

1 Introduction

An absolute valued algebra (AVA) is a non-zero algebra $\mathcal{A}$ over the field $\mathbb{R}$ of real numbers endowed with a norm $||.||$ satisfying

$$||xy|| = ||x||\ ||y|| \text{ for all } x, y \in \mathcal{A}. \tag{1}$$

The most natural absolute-valued algebras are $\mathbb{R}$, $\mathbb{C}$, $\mathbb{H}$ (Hamilton's quaternions) and $\mathbb{O}$ (Cayley's octonions).

It is well known that the dimension of a finite-dimensional AVA is either 1, 2, 4, or 8 and that the norm of the algebra comes from an inner product [1]. The reader is referred to [15] for a survey of studies, within the theory, prior to 2004.

The degree of a finite-dimensional algebra $\mathcal{A}$ is the smallest natural number n such that all single-generated subalgebras of $\mathcal{A}$ have dimension $\leq n$ [14]. It follows from [1] that finite-dimensional AVAs have degree 1, 2, 4, or 8.

P. A. Dieng · A. Diouf (✉)
Département de Mathématiques et Informatique, Faculté des Sciences et Techniques, Université Cheikh Anta Diop, Dakar, Senegal
e-mail: alassane1.diouf@ucad.edu.sn

A. Rochdi
Département de Mathématiques et Informatique, Faculté des Sciences Ben M'Sik, Université Hassan II, Casablanca, Morocco

© The Author(s), under exclusive license to Springer Nature Switzerland AG 2026
C. T. Gueye et al. (eds.), *Mathematics, Computer Science, Cybersecurity and Artificial Intelligence*, Springer Proceedings in Mathematics & Statistics 536,
https://doi.org/10.1007/978-3-032-17666-0_1

Infinite dimensional AVAs exist even in presence of left unit [5, 13], or satisfaction of the identity $(x^2, y, x^2) = 0$ [9], [6, Proposition 2].

Four-dimensional AVAs are the so-called *principal isotopes* of the algebra of quaternions $\mathbb{H}$, namely $\mathbb{H}(a, b)$, ${}^*\mathbb{H}(a, b)$, $\mathbb{H}^*(a, b)$, $\overset{*}{\mathbb{H}}(a, b)$ obtained from fixed norm-one elements a, b in $\mathbb{H}$ by endowing the normed space of $\mathbb{H}$ with the products

$$\begin{aligned} x \diamond_1 y &= axyb, \\ x \diamond_2 y &= \overline{x}ayb, \\ x \diamond_3 y &= axb\overline{y}, \\ x \diamond_4 y &= a\overline{x}\,\overline{y}b, \end{aligned}$$

respectively [11, page 170]. Among these algebras, those with left-unit are precisely $\mathbb{H}(a, 1)$, ${}^*\mathbb{H}(a, 1)$ [11, Corollary 3.1].

We have the following eight-dimensional AVA with left-unit:

$$(\mathbb{O}, \odot) := {}^*\mathbb{O}(i, 1), \text{ with } x \odot y = (\overline{x}i)y \text{ having left-unit } e = i.$$

We define, from an arbitrary four-dimensional AVA, the following process which allows one to construct a family of eight-dimensional AVAs. Let Φ, Ψ be linear isometries of the Euclidian space $\mathbb{H}$, fixing 1, and let $*$ be the Cayley-Dickson product in $\mathbb{H} \times \mathbb{H}$. For arbitrary linear isometries φ, ψ, of $\mathbb{H}$ we define on the space $\mathbb{H} \times \mathbb{H}$ the product

$$(x, y) \diamond (u, v) = \big(\Phi(x), \varphi(y)\big) * \big(\Psi(u), \psi(v)\big).$$

We obtain an eight-dimensional AVA $\mathbb{H} \times \mathbb{H}_{(\Phi,\varphi),(\Psi,\psi)}$, where (Φ, φ) denotes the linear isometry of $\mathbb{H} \times \mathbb{H}$ such that $(\Phi, \varphi)(x, y) = \big(\Phi(x), \varphi(y)\big)$. Moreover, the subset $\{(x, 0) : x \in \mathbb{H}\}$ of $\mathbb{H} \times \mathbb{H}_{(\Phi,\varphi),(\Psi,\psi)}$ is a subalgebra isomorphic to $\mathbb{H}_{\Phi,\Psi}$. We say that $\mathbb{H} \times \mathbb{H}_{(\Phi,\varphi),(\Psi,\psi)}$ is obtained from the algebra $\mathbb{H}$ by *duplication process* [2].

Eight-dimensional AVAs with left-unit obtained by the duplication process are of the form $\mathbb{H} \times \mathbb{H}_{(\Phi,\varphi),(I_{\mathbb{H}},I_{\mathbb{H}})}$ [12] which we simply note $\mathbb{H} \times \mathbb{H}_{(\Phi,\varphi)}$. Their left-unit being $e = (1, 0)$.

Let Φ_1, Φ_2, φ_1, φ_2 be linear isometries of the Euclidian space $\mathbb{H}$ with $\Phi_1(1) = \Phi_2(1) = 1$. We assume that the algebras $\mathbb{H} \times \mathbb{H}_{(\Phi_1,\varphi_1)}$, $\mathbb{H} \times \mathbb{H}_{(\Phi_2,\varphi_2)}$ are isomorphic. They are said to be *naturally isomorphic* if they are linked by an isomorphism of the form (f, g) where f, g are, necessarily, isometries of the Euclidean space $\mathbb{H}$ with $f(1) = 1$. They are said to be *exceptionally isomorphic* otherwise.

This present work is a revision of both [3, 7]. We show that every AVA $\mathcal{A}$ with left-unit satisfying $(x^2, x^2, x^r) = 0$ for fixed $r \in \{1, 2\}$ is finite-dimensional (Lemma 1). Moreover, for every AVA $\mathcal{A}$ with left-unit, the identity $(x^2, x^2, x) = 0$ is equivalent to $(x^2, x^2, x^2) = 0$ (Proposition 2). In these conditions $\mathcal{A}$ has degree ≤ 4 (Proposition 2). Theorem 1 proves that the algebras ${}^*\mathbb{H}(i, 1)$, ${}^*\mathbb{O}(i, 1)$ and $\widetilde{\mathbb{O}}$ have been forgotten in [3, Proposition 4.8 and Theorem 4.10]. We also specify that the algebra $\widetilde{\mathbb{O}}\ (i) := \mathbb{H} \times \mathbb{H}_{(T_{i,\bar{i}} \circ \sigma_{\mathbb{H}}, T_{i,i})}$ in [7, Theorem 5] does not appear in the Theorem 1, because $\widetilde{\mathbb{O}}\ (i)$ is exceptionally isomorphic $\widetilde{\mathbb{O}} := \mathbb{H} \times \mathbb{H}_{(\sigma_{\mathbb{H}}, I_{\mathbb{H}})}$ (Lemma 4). Next, we get a reduced list of all AVAs $\mathcal{A}$ with left unit satisfying $(x^2, x^2, x^r) = 0$:

$\deg(\mathcal{A})$	Reduced list of isomorphism classes
≤ 2	$\mathbb{R}$, $\mathbb{C}$, $^*\mathbb{C}$, $\mathbb{H}$, $^*\mathbb{H}$, $\mathbb{O}$, $^*\mathbb{O}$
4	$^*\mathbb{H}(i, 1)$, $^*\mathbb{O}(i, 1)$, $\widetilde{\mathbb{O}}$

2 Notations and Definitions

Let $\mathcal{A}$ be an algebra over the field $\mathbb{R}$ of real numbers and $\Phi, \Psi : \mathcal{A} \to \mathcal{A}$ be two linear bijections. We denote by $\mathcal{A}_{\Phi,\Psi}$ the algebra having underlying vector space $\mathcal{A}$ and product $a \diamond b = \Phi(a)\Psi(b)$ and we put $\mathcal{A}_{\Phi,\Psi} := (\mathcal{A}, \diamond)$. It is said to be *isotopic* to $\mathcal{A}$ and obtained from $\mathcal{A}$ and $\{\Phi, \Psi\}$ by *isotopy*. If $I_{\mathcal{A}}$ is the identity mapping, then algebra $\mathcal{A}_{\Phi, I_{\mathcal{A}}}$ is denoted $\mathcal{A}_{\Phi}$.

Let now $\mathbb{A} \in \{\mathbb{C}, \mathbb{H}, \mathbb{O}\}$ and f, g be linear isometries of the Euclidean space $\mathbb{A}$, $\mathbb{A}_{f,g}$ is an absolute-valued algebra.

Let A be an absolute-valued algebra whose norm comes from the inner product $\langle ., . \rangle$, then

$$\langle yx, zx \rangle = ||x||^2 \langle y, z \rangle \ \text{ for all } x, y, z \in \mathcal{A}. \tag{2}$$

Let $a, b, c \in \mathcal{A}$. We denote by L_a, R_a, $T_{a,b}$, (a, b, c), $[a, b]$, $\mathbb{S}(E)$, respectively, the operators of multiplication by a on the left $x \mapsto ax$, on the right $x \mapsto xa$, the operator $L_a \circ R_b$, the associator $(ab)c - a(bc)$ of a, b, c, the commutator $ab - ba$ of a, b, and the unit sphere of a normed space E.

Note 1 Let $\mathbb{A}$ be one of the principal absolute-valued algebras $\mathbb{R}$, $\mathbb{C}$, $\mathbb{H}$, $\mathbb{O}$ provided with its standard involution $\sigma_{\mathbb{A}} : x \mapsto \overline{x}$ and absolute value $|x| = \sqrt{x\overline{x}}$. Clearly, $\mathbb{A}$ and its standard isotope $\mathbb{A}_{\sigma_{\mathbb{A}}, I_{\mathbb{A}}} = {}^*\mathbb{A}$ are finite-dimensional AVAs with left-unit 1 and satisfy $(x^2, x^2, x^2) = 0$.

Let now $r \in \{1, 2\}$. There are mappings $f_1, \ldots, f_6 : \mathcal{A} \times \mathcal{A} \to \mathcal{A}$ for which the following equality holds for all $x, y \in \mathcal{A}$ and $\lambda \in \mathbb{R}$:

$$\big((x + \lambda y)^2, (x + \lambda y)^2, (x + \lambda y)^r\big) = (x^2, x^2, x^r) + \sum_{h=1}^{4+r} \lambda^h f_h(x, y). \tag{3}$$

The identity $(x^2, x^2, x^r) = 0$ in $\mathcal{A}$ is equivalent to $f_m = 0\ (1 \leq m \leq 4 + r)$ called the mth identity obtained from $(x^2, x^2, x^r) = 0$ by linearization. This identity is denoted by $(r.m)$ and is specified, partially, in the following table where $x \bullet y = xy + yx$:

Table 1 Linearizations

(r, m)	$(r.m)$
$(1, 1)$	$(x^2, x^2, y) + (x^2, x \bullet y, x) + (x \bullet y, x^2, x) = 0$
$(2, 1)$	$(x^2, x^2, x \bullet y) + (x^2, x \bullet y, x^2) + (x \bullet y, x^2, x^2) = 0$
$(1, 2)$	$(x^2, x \bullet y, y) + (x^2, y^2, x) + (x \bullet y, x^2, y) + (x \bullet y, x \bullet y, x) + (y^2, x^2, x) = 0$
$(2, 2)$	$(x^2, x^2, y^2) + (x^2, x \bullet y, x \bullet y) + (x^2, y^2, x^2) + (x \bullet y, x^2, x \bullet y) + (x \bullet y, x \bullet y, x^2) + (y^2, x^2, x^2) = 0$

3 Preliminary Results

In this section $\mathcal{A}$ is assumed to be an AVA with left-unit e.

It is well known that the norm $||.||$ of $\mathcal{A}$ comes from an inner product $\langle ., . \rangle$ [13, Remark 4. (i)] and the following equalities hold for all $x, y, z \in \mathcal{A}$ with $x^* = 2\langle e, x\rangle e - x$ [15, Theorem 3.5]:

$$\langle xy, z\rangle = \langle y, x^* z\rangle, \tag{4}$$

$$x^*(xy) = ||x||^2 y. \tag{5}$$

By putting $y = e$ in (3.5) we get

$$x(xe) = 2\langle e, x\rangle xe - ||x||^2 e \text{ for all } x \in \mathcal{A}. \tag{6}$$

We can now state the following preliminary results:

Lemma 1 *If $\mathcal{A}$ satisfies $(x^2, x^2, x^r) = 0$ for fixed $r \in \{1, 2\}$ then $R_e^2 = I_{\mathcal{A}}$ and $\mathcal{A}$ is finite-dimensional. Moreover, the four equalities hold for all $y, z \in \mathcal{A}$:*

(1) $(y + ye, e, z) = 0$,
(2) $\big((ze)z\big)e = z(ze)$,
(3) $\big(z(ze)\big)e = (ze)z$,
(4) $y^2 e = y^2$.

Proof By putting $x = e$ in Table 1 for any identity of the form $(r.1)$ we get $(y + ye, e, e) = 0$. Now

$$\begin{aligned} 0 &= (y + ye, e, e) \\ &= \big((y + ye)e - (y + ye)\big)e \\ &= \big((ye)e - y\big)e. \end{aligned}$$

We deduce that $(ye)e = y$ because $\mathcal{A}$ has no divisors of zero, that is $R_e^2 = I_{\mathcal{A}}$. So L_e, R_e are both invertible and then $\mathcal{A}$ is finite-dimensional [15, Theorem 2.2].

Now let $x, y, z \in \mathcal{A}$.

(1) We have

$$\begin{aligned} (y + ye, e, z) &= \big((y + ye)e - (y + ye)\big)z \\ &= \big((ye)e - y\big)z \\ &= 0. \end{aligned}$$

(2) We have

$$\begin{aligned} (ze)z &= (ze)\big((ze)e\big) \\ &= 2\langle e, ze\rangle(ze)e - ||ze||^2 e \quad \text{by putting } x = ze \text{ in equality (3.6)} \\ &= 2||e||^2\langle e, z\rangle(ze)e - ||ze||^2 e \quad \text{by equality (2.2)} \\ &= 2\langle e, z\rangle z - ||z||^2 e \quad \text{because } ||e|| = 1 \text{ and } R_e^2 = I_{\mathcal{A}}. \end{aligned}$$

Consequently

$$\begin{aligned} \big((ze)z\big)e &= 2\langle e, z\rangle ze - ||z||^2 e \\ &= z(ze) \quad \text{from equality (3.6)}. \end{aligned}$$

(3) We have

$$\begin{aligned} \big(z(ze)\big)e &= R_e^2(\big((ze)z\big) \quad \text{according to the previous equality (2)} \\ &= (ze)z. \end{aligned}$$

(4) By putting $x = e$ in each of equalities (1.2), (2), we get

$$(y + ye, y + ye, e) + (y^2, e, e) = 0.$$

Now

$$
\begin{aligned}
0 &= (y + ye, y + ye, e) + (y^2, e, e) \\
&= \big((y + ye)^2 e - (y + ye)^2\big) + (y^2 e - y^2)e \;\text{ because } R_e^2 = I_{\mathcal{A}} \\
&= \big(y(ye) + (ye)y + (ye)^2\big)e - \big(y(ye) + (ye)y + (ye)^2\big) \;\text{ because } R_e^2 = I_{\mathcal{A}} \\
&= \big((ye)y + (ye)^2\big)e - \big(y(ye) + (ye)^2\big) \;\text{ according to the previous equality (3)} \\
&= (ye)^2 e - (ye)^2 \;\text{ according to the previous equality (2).}
\end{aligned}
$$

So $y^2 e = y^2$ because R_e is bijective.

□

It is well known that $\mathbb{R}$ and the standard isotopes $\mathbb{C}$, ${}^*\mathbb{C}$, $\mathbb{C}^*$, $\overset{*}{\mathbb{C}}$ of $\mathbb{C}$ are the only AVAs of dimension ≤ 2 [14, Lemma 2]. Among these algebras only $\mathbb{R}$, $\mathbb{C}$, ${}^*\mathbb{C}$ contain a left unit $e = 1$ and satisfy either $(x^2, x^2, x) = 0$ or $(x^2, x^2, x^2) = 0$.

We have the following classification for algebras of dimension ≤ 4.

Proposition 1 *Let $\mathcal{A}$ be an AVA of dimension ≤ 4 with left unit. Then the following three statements are equivalent:*

(1) $\mathcal{A}$ satisfies $(x^2, x^2, x) = 0$,
(2) $\mathcal{A}$ satisfies $(x^2, x^2, x^2) = 0$,
(3) $\mathcal{A}$ is isomorphic to either $\mathbb{R}$, $\mathbb{C}$, ${}^*\mathbb{C}$, $\mathbb{H}$, ${}^*\mathbb{H}$ or ${}^*\mathbb{H}(i, 1)$.

Proof The theorem is clear in dimension ≤ 2. Now, the unique four-dimensional AVAs with left unit e are the principal isotopes of $\mathbb{H}$ of the form $\mathbb{H}(a, 1)$ (with $e = \overline{a}$) and ${}^*\mathbb{H}(a, 1)$ (with $e = a$) for some norm-one $a \in \mathbb{H}$ [11, Corollary 3.1]. Those who satisfy $(x^2, x^2, x) = 0$ are the same as those which satisfy $(x^2, x^2, x^2) = 0$, and are isomorphic to either $\mathbb{H}$, ${}^*\mathbb{H}$ or ${}^*\mathbb{H}(i, 1)$ [4, Theorem 3.24]. □

We obtain the following result.

Proposition 2 *Let $\mathcal{A}$ be an eight-dimensional AVA with left unit. The three following statements are equivalent:*

(1) $\mathcal{A}$ satisfies $(y^2, y^2, y) = 0$,
(2) $y^2 e = y^2$ for all $y \in \mathcal{A}$,
(3) $\mathcal{A}$ satisfies $(y^2, y^2, y^2) = 0$.

Under these conditions $\mathcal{A}$ has degree ≤ 4.

Proof Implications (1) $\Rightarrow$ (2), (3) $\Rightarrow$ (2) are given by Lemma 1, and implication (2) $\Rightarrow$ (3) holds by putting $x = y^2$ in equality (6) taking into account that $(y^2)^2 = 2\langle e, y^2\rangle y^2 - ||y^2||^2 e$ commutes with y^2.

Now, if $y^2 e = y^2$ for all $y \in \mathcal{A}$ then $\mathcal{A}$ has degree ≤ 4 ([7, Theorem 3], [6, Corollary 5]). We deduce that $\mathcal{A}$ satisfies $(y^2, y^2, y) = 0$ by Proposition 1. This establishes the implication (2) $\Rightarrow$ (1) and completes the proof. □

In what follows we treat the case of algebras of dimension 8.

4 Two Examples for Algebras of Degree Four

For every $x, y \in \mathbb{O}$, the fact that $x, y, \overline{x}, \overline{y}$ belong to an associative subalgebra of $\mathbb{O}$ [16, Theorem 3.1] will be used in the sequel without further reference. Since $\mathbb{O}$ is alternative, then $\mathbb{O}$ satisfies the Moufang identities.

4.1 The Algebra $^*\mathbb{O}(i, 1)$

Let $b \in \mathbb{S}\big(Im(\mathbb{O})\big)$. We have the following eight-dimensional AVA with left-unit:

$$(\mathbb{O}, \overset{b}{\diamond}) := \mathbb{O}_{T_{b,\overline{b}} \circ \sigma_{\mathbb{O}}}, \text{ with } x \overset{b}{\diamond} y = (b\overline{x}\,\overline{b})y \text{ having left-unit } e = 1.$$

Proposition 3 *The algebras* $^*\mathbb{O}(i, 1)$, $\mathbb{O}_{T_{b,\overline{b}} \circ \sigma_{\mathbb{O}}}$ *are isomorphic and satisfy* $(z^2, z^2, z^2) = 0$.

Proof According to [10, Lemme 1, pp. 269–270] there exists an automorphism Ψ of the algebra $\mathbb{O}$ such that $\Psi(b) = i$. For every $x, y \in \mathbb{O}$, we have

$$\begin{aligned}
x \overset{b}{\diamond} y &= -(b\overline{x}b)y. \\
L_b(x \overset{b}{\diamond} y) &= -b\big((b\overline{x}b)y\big) \\
&= -b\Big(b\big(\overline{x}(by)\big)\Big) \text{ by left-Moufang identity.} \\
&= \overline{x}(by). \\
(\Psi \circ L_b)(x \overset{b}{\diamond} y) &= \Psi\big(\overline{x}(by)\big) \\
&= \Psi(\overline{x})\big(\Psi(b)\Psi(y)\big) \\
&= \overline{\Psi(x)}\big(i\Psi(y)\big), \\
(\Psi \circ L_b)(x) \odot (\Psi \circ L_b)(y) &= \Psi(bx) \odot \Psi(by) \\
&= \big(\overline{\Psi(bx)}\, i\big)\Psi(by) \\
&= \big(\overline{\Psi(b)\Psi(x)}\, i\big)\big(\Psi(b)\Psi(y)\big) \\
&= \big(\overline{\Psi(x)}\, \overline{i}\, i\big)\big(i\Psi(y)\big) \\
&= \overline{\Psi(x)}\big(i\Psi(y)\big).
\end{aligned}$$

So $\Psi \circ L_b : \mathbb{O}_{T_{b,\overline{b}} \circ \sigma_{\mathbb{O}}} \to {}^*\mathbb{O}(i, 1)$ is an isomorphism of algebras. Finally

$$\begin{aligned}(x \odot x) \odot i &= (\overline{x}ix) \odot i \\ &= \big(\overline{\overline{x}ix}\, i\big)i \\ &= \overline{x}\,\overline{i}x\, i^2 \\ &= \overline{x}ix \\ &= x \odot x.\end{aligned}$$

Proposition 2 concludes the proof. □

4.2 The Algebra $\tilde{\mathbb{O}}$

We need the following preliminary results.

Lemma 2 *For every norm-one $u \in \mathbb{H}$, the mappings $(T_{u,\overline{u}}, T_{u,\overline{u}}) := \Lambda_u$, $(I_{\mathbb{H}}, L_u) := \omega_u$ are automorphisms of the octonion algebra $(\mathbb{H} \times \mathbb{H}, *)$ and, clearly, ${\Lambda_u}^{-1} = \Lambda_{\overline{u}}$, ${\omega_u}^{-1} = \omega_{\overline{u}}$.*

Proof Let $(x, y), (x', y') \in \mathbb{H} \times \mathbb{H}$. We have:

$$\begin{aligned}\Lambda_u(x, y) * \Lambda_u(x', y') &= (ux\overline{u}, uy\overline{u}) * (ux'\overline{u}, uy'\overline{u}) \\ &= (ux\overline{u}.ux'\overline{u} - \overline{uy'\overline{u}}.uy\overline{u}, uy\overline{u}.\overline{ux'\overline{u}} + uy'\overline{u}.ux\overline{u}) \\ &= (uxx'\overline{u} - u\overline{y'}\,\overline{u}.uy\overline{u}, uy\overline{u}.u\overline{x'}\,\overline{u} + uy'x\overline{u}) \\ &= \big(u(xx' - \overline{y'}y)\overline{u}, u(y\overline{x'} + y'x)\overline{u}\big) \\ &= \Lambda_u(xx' - \overline{y'}y, y\overline{x'} + y'x) \\ &= \Lambda_u\big((x, y) * (x', y')\big),\end{aligned}$$

and

$$\begin{aligned}\omega(x, y) * \omega(x', y') &= (x, uy) * (x', uy') \\ &= (xx' - \overline{uy'}.uy, uy.\overline{x'} + uy'.x) \\ &= \big(xx' - \overline{y'}y, u(y\overline{x'} + y'x)\big) \\ &= \omega\big((x, y) * (x', y')\big).\end{aligned}$$

which completes the proof. □

Lemma 3 *Let $(a, b) \in \mathbb{S}\big(Im(\mathbb{H})\big) \times \mathbb{S}(\mathbb{H})$. Then*

(1) The algebra $\mathbb{H} \times \mathbb{H}_{(T_{a,\overline{a}} \circ \sigma_{\mathbb{H}}, T_{b,a})}$ is naturally isomorphic to $\mathbb{H} \times \mathbb{H}_{(T_{i,\overline{i}} \circ \sigma_{\mathbb{H}}, T_{ub\overline{u},i})}$.
(2) If, moreover, $b \in \mathbb{S}\big(Im(\mathbb{H})\big)$ then the algebra $\mathbb{H} \times \mathbb{H}_{(T_{a,\overline{a}} \circ \sigma_{\mathbb{H}}, T_{b,a})}$ is naturally isomorphic to $\mathbb{H} \times \mathbb{H}_{(T_{i,\overline{i}} \circ \sigma_{\mathbb{H}}, T_{i,i})}$.

Proof There is a norm-one $u \in \mathbb{H}$ such that $ua\overline{u} = i$ [4, Lemma 2.10]. Note also that $T_{u,\overline{u}}$ commutes with $\sigma_{\mathbb{H}}$.

(1) We have

$$\begin{aligned}(T_{u,\overline{u}}, T_{u,\overline{u}}) \circ (T_{a,\overline{a}} \circ \sigma_{\mathbb{H}}, T_{b,a}) \circ (T_{\overline{u},u}, T_{\overline{u},u}) &= (T_{u,\overline{u}} \circ T_{a,\overline{a}} \circ \sigma_{\mathbb{H}} \circ T_{\overline{u},u}, T_{u,\overline{u}} \circ T_{b,a} \circ T_{\overline{u},u}) \\ &= (T_{u,\overline{u}} \circ T_{a,\overline{a}} \circ T_{\overline{u},u} \circ \sigma_{\mathbb{H}}, T_{u,\overline{u}} \circ T_{b,a} \circ T_{\overline{u},u}) \\ &= (T_{ua\overline{u},u\overline{a}\,\overline{u}} \circ \sigma_{\mathbb{H}}, T_{ub\overline{u},ua\overline{u}}) \\ &= (T_{i,\overline{i}} \circ \sigma_{\mathbb{H}}, T_{ub\overline{u},i}).\end{aligned}$$

Lemma 2 and [12, Theorem 4.3] conclude the first statement.

(2) We have $ub\overline{u} \in \mathbb{S}(Im(\mathbb{H}))$ and there is a norm-one $v \in \mathbb{H}$ such that $v(ub\overline{u})\overline{v} = i$. Now

$$\begin{aligned}(I_{\mathbb{H}}, L_v) \circ (T_{i,\overline{i}} \circ \sigma_{\mathbb{H}}, T_{ub\overline{u},i}) \circ (I_{\mathbb{H}}, l_{\overline{v}}) &= (T_{i,\overline{i}} \circ \sigma_{\mathbb{H}}, L_v \circ L_{ub\overline{u}} \circ R_i \circ L_{\overline{v}}) \\ &= (T_{i,\overline{i}} \circ \sigma_{\mathbb{H}}, L_v \circ L_{ub\overline{u}} \circ L_{\overline{v}} \circ R_i) \\ &= (T_{i,\overline{i}} \circ \sigma_{\mathbb{H}}, L_{v(ub\overline{u})\overline{v}} \circ R_i) \\ &= (T_{i,\overline{i}} \circ \sigma_{\mathbb{H}}, L_i \circ R_i) \\ &= (T_{i,\overline{i}} \circ \sigma_{\mathbb{H}}, T_{i,i}).\end{aligned}$$

Lemma 2 and [12, Theorem 4.3] conclude the second statement.

□

Lemma 4 *The algebra $\mathbb{H} \times \mathbb{H}_{(\sigma_{\mathbb{H}}, I_{\mathbb{H}})}$ satisfies $(x^2, x^2, x^2) = 0$ and is exceptionally isomorphic to $\mathbb{H} \times \mathbb{H}_{(T_{i,\overline{i}} \circ \sigma_{\mathbb{H}}, T_{i,i})}$.*

Proof Let $\diamond$, $\odot$ be the products of the algebras $\mathbb{H} \times \mathbb{H}_{(\sigma_{\mathbb{H}}, I_{\mathbb{H}})}$ and $\mathbb{H} \times \mathbb{H}_{(T_{i,\overline{i}} \circ \sigma_{\mathbb{H}}, T_{i,i})}$, respectively, and let $(y, z) \in \mathbb{H} \times \mathbb{H}$. We have

$$\begin{aligned}\big((y, z) \diamond (y, z)\big) \diamond (1, 0) &= \big((\overline{y}, z) * (y, z)\big) \diamond (1, 0) \\ &= (|y|^2 - |z|^2, 2z\overline{y}) \diamond (1, 0) \\ &= (|y|^2 - |z|^2, 2z\overline{y}) * (1, 0) \\ &= (|x|^2 - |y|^2, 2z\overline{y}) \\ &= (y, z) \diamond (y, z).\end{aligned}$$

Proposition 2 concludes the first statement.

Suppose now that there exists an isomorphism

$$(f, g) : \mathbb{H} \times \mathbb{H}_{(\sigma_{\mathbb{H}}, I_{\mathbb{H}})} \to \mathbb{H} \times \mathbb{H}_{(T_{i,\overline{i}} \circ \sigma_{\mathbb{H}}, T_{i,i})}$$

where f, g are, necessarily, linear isometries of the Euclidean space $\mathbb{H}$ with $(f, g)(1, 0) = (1, 0)$. Let $1, i, j, k$ be the canonical basis of space $\mathbb{H}$. We have

$$
\begin{aligned}
(f,g)\Big(\big(f^{-1}(j),0\big)\diamond\big(f^{-1}(j),0\big)\Big) &= (f,g)\Big(\big(\overline{f^{-1}(j)},0\big)*\big(f^{-1}(j),0\big)\Big)\\
&= (f,g)\big(|f^{-1}(j)|^2,0\big)\\
&= (f,g)(|j|^2,0) \text{ because } f^{-1} \text{ is an isometry}\\
&= (f,g)(1,0)\\
&= (1,0).
\end{aligned}
$$

On the other hand

$$
\begin{aligned}
(f,g)\big(f^{-1}(j),0\big)\odot(f,g)\big(f^{-1}(j),0\big)\big) &= (j,0)\odot(j,0)\\
&= (i\bar{j}\,\bar{i},0)*(j,0)\\
&= (i\bar{j}\,\bar{i}j,0)\\
&= (ji\bar{i}j,0)\\
&= (-1,0).
\end{aligned}
$$

Absurd, so $\mathbb{H}\times\mathbb{H}_{(\sigma_{\mathbb{H}},I_{\mathbb{H}})}$, $\mathbb{H}\times\mathbb{H}_{(T_{i,\bar{i}}\circ\sigma_{\mathbb{H}},T_{i,i})}$ are not naturally isomorphic.

We will show that these algebras are isomorphic using multiplication tables. Indeed, the multiplication tables of algebras

$$
\mathbb{H}\times\mathbb{H}_{(\sigma_H,I_{\mathbb{H}})} := (\mathbb{H}\times\mathbb{H},\diamond),\quad \mathbb{H}\times\mathbb{H}_{(T_{i,\bar{i}}\circ\sigma_{\mathbb{H}},T_{i,i})} := (\mathbb{H}\times\mathbb{H},\odot)
$$

with respect to the canonical basis

$$
e_0=(1,0),\ e_1=(i,0),\ e_2=(j,0),\ e_3=(k,0),\ e_4=(0,1),\ e_5=(0,i),\ e_6=(0,j),\ e_7=(0,k)
$$

of the vector space $\mathbb{H}\times\mathbb{H}$ are given, respectively, by

Table 2 First product table in algebra

$\diamond$	e_0	e_1	e_2	e_3	e_4	e_5	e_6	e_7
e_0	e_0	e_1	e_2	e_3	e_4	e_5	e_6	e_7
e_1	$-e_1$	e_0	$-e_3$	e_2	$-e_5$	e_4	e_7	$-e_6$
e_2	$-e_2$	e_3	e_0	$-e_1$	$-e_6$	$-e_7$	e_4	e_5
e_3	$-e_3$	$-e_2$	e_1	e_0	$-e_7$	e_6	$-e_5$	e_4
e_4	e_4	$-e_5$	$-e_6$	$-e_7$	$-e_0$	e_1	e_2	e_3
e_5	e_5	e_4	$-e_7$	e_6	$-e_1$	$-e_0$	$-e_3$	e_2
e_6	e_6	e_7	e_4	$-e_5$	$-e_2$	e_3	$-e_0$	$-e_1$
e_7	e_7	$-e_6$	e_5	e_4	$-e_3$	$-e_2$	e_1	$-e_0$

Table 3 Second product table in algebra

$\odot$	e_0	e_1	e_2	e_3	e_4	e_5	e_6	e_7
e_0	e_0	e_1	e_2	e_3	e_4	e_5	e_6	e_7
e_1	$-e_1$	e_0	$-e_3$	e_2	$-e_5$	e_4	e_7	$-e_6$
e_2	e_2	$-e_3$	$-e_0$	e_1	e_6	e_7	$-e_4$	$-e_5$
e_3	e_3	e_2	$-e_1$	$-e_0$	e_7	$-e_6$	e_5	$-e_4$
e_4	$-e_4$	e_5	e_6	e_7	e_0	$-e_1$	$-e_2$	$-e_3$
e_5	$-e_5$	$-e_4$	e_7	$-e_6$	e_1	e_0	e_3	$-e_2$
e_6	e_6	e_7	e_4	$-e_5$	$-e_2$	e_3	$-e_0$	$-e_1$
e_7	e_7	$-e_6$	e_5	e_4	$-e_3$	$-e_2$	e_1	$-e_0$

Now, the multiplication table of algebra $(\mathbb{H} \times \mathbb{H}, \odot)$ with respect to the basis

$$e'_0 = (1, 0),\ e'_1 = (i, 0),\ e'_2 = (0, 1),\ e'_3 = (0, i),\ e'_4 = (0, -j),\ e'_5 = (0, k),\ e'_6 = (-j, 0),\ e'_7 = (k, 0)$$

is given by

Table 4 Third product table in algebra

$\odot$	e'_0	e'_1	e'_2	e'_3	e'_4	e'_5	e'_6	e'_7
e'_0	e'_0	e'_1	e'_2	e'_3	e'_4	e'_5	e'_6	e'_7
e'_1	$-e'_1$	e'_0	$-e'_3$	e'_2	$-e'_5$	e'_4	e'_7	$-e'_6$
e'_2	$-e'_2$	e'_3	e'_0	$-e'_1$	$-e'_6$	$-e'_7$	e'_4	e'_5
e'_3	$-e'_3$	$-e'_2$	e'_1	e'_0	$-e'_7$	e'_6	$-e'_5$	e'_4
e'_4	e'_4	$-e'_5$	$-e'_6$	$-e'_7$	$-e'_0$	e'_1	e'_2	e'_3
e'_5	e'_5	e'_4	$-e'_7$	e'_6	$-e'_1$	$-e'_0$	$-e'_3$	e'_2
e'_6	e'_6	e'_7	e'_4	$-e'_5$	$-e'_2$	e'_3	$-e'_0$	$-e'_1$
e'_7	e'_7	$-e'_6$	e'_5	e'_4	$-e'_3$	$-e'_2$	e'_1	$-e'_0$

We see that Tables 2 and 4 are identical. Thus, these two algebras are isomorphic. We will subsequently note $\widetilde{\mathbb{O}}$ the algebra $\mathbb{H} \times \mathbb{H}_{(\sigma_{\mathbb{H}}, I_{\mathbb{H}})}$ (Tables 3 and 5). $\square$

Let us denote $\mathcal{A}_e$ the space $\{x \in \mathcal{A} : xe = x\}$. Clearly, if $\psi : \mathcal{A} \to \mathcal{B}$ is an isomorphism of algebras then $\mathcal{B}$ contains a left-unit $\psi(e)$ and we have $\psi(\mathcal{A}_e) = \mathcal{B}_{\psi(e)}$. If, moreover, $\mathcal{A}$ is finite-dimensional then $\mathcal{B}$ is finite-dimensional and we have $\dim \mathcal{B}_{\psi(e)} = \dim \mathcal{A}_e$.

The following table illustrates the subspaces $\mathcal{A}_e$ and their dimensions for the algebras $\mathbb{O}$, $^*\mathbb{O}$, $^*\mathbb{O}(i, 1)$, $\widetilde{\mathbb{O}}$.

Table 5 Dimension of the subspace Ae

Algebra $\mathcal{A}$	Subspace $\mathcal{A}_e$	dim $\mathcal{A}_e$
$\mathbb{O}$	$\mathbb{O}$	8
${}^*\mathbb{O}$	$\mathbb{R}$	1
${}^*\mathbb{O}(i,1)$	$Im(\mathbb{O})$	7
$\widetilde{\mathbb{O}}$	$\mathbb{R}\times\mathbb{H}$	5

Proposition 4 *The four algebras* $\mathbb{O}$, ${}^*\mathbb{O}$, ${}^*\mathbb{O}(i,1)$, $\widetilde{\mathbb{O}}$ *are mutually non-isomorphic.*

5 When Do the Algebras $\mathbb{H}\times\mathbb{H}_{(\Phi,\Psi)}$ Satisfy $(x^2,x^2,x^2)=0$?

Let $\mathcal{O},\mathcal{O}^+,\mathcal{O}^-$ be, respectively, the orthogonal group of all linear isometries of Euclidean space $\mathbb{H}$, its subgroup of proper linear isometries and its subset of improper linear isometries. Clearly, L_a, R_b, $T_{a,b}$ belong to $\mathcal{O}$ for any norm-one $a,b\in\mathbb{H}$. We will also use the following notations:

$$\begin{aligned}
\mathcal{I}^+ &= \{f\in\mathcal{O}^+ : f \text{ involutive}\},\\
\mathcal{I}^- &= \{f\in\mathcal{O}^- : f \text{ involutive}\},\\
\mathcal{P}_1 &= \{f\in\mathcal{P} : f(1)=1\} \text{ for all } \mathcal{P}\in\{\mathcal{O}^+,\mathcal{O}^-,\mathcal{I}^+,\mathcal{I}^-\}.
\end{aligned}$$

Lemma 5 $\mathcal{O}^+,\mathcal{O}^-,\mathcal{O}_1^+,\mathcal{O}_1^-,\mathcal{I}^+,\mathcal{I}^-,\mathcal{I}_1^+,\mathcal{I}_1^-$ *are given by:*

$$\begin{aligned}
\mathcal{O}^+ &= \{T_{a,b} : a,b\in\mathbb{S}(\mathbb{H})\}, &(7)\\
\mathcal{O}^- &= \{T_{a,b}\circ\sigma_{\mathbb{H}} : a,b\in\mathbb{S}(\mathbb{H})\} := \mathcal{O}^+\circ\sigma_{\mathbb{H}}, &(8)\\
\mathcal{O}_1^+ &= \{T_{a,\overline{a}} : a\in\mathbb{S}(\mathbb{H})\}, &(9)\\
\mathcal{O}_1^- &= \{T_{a,\overline{a}}\circ\sigma_{\mathbb{H}} : a\in\mathbb{S}(\mathbb{H})\} := \mathcal{O}_1^+\circ\sigma_{\mathbb{H}}, &(10)\\
\mathcal{I}^+ &= \{\pm I_{\mathbb{H}}\}\cup\{T_{a,b} : a,b\in\mathbb{S}\big(Im(\mathbb{H})\big)\}, &(11)\\
\mathcal{I}^- &= \{\pm T_{a,a}\circ\sigma_{\mathbb{H}} : a\in\mathbb{S}(\mathbb{H})\}, &(12)\\
\mathcal{I}_1^+ &= \{I_{\mathbb{H}}\}\cup\{T_{a,\overline{a}} : a\in\mathbb{S}\big(Im(\mathbb{H})\big)\}, &(13)\\
\mathcal{I}_1^- &= \{\sigma_{\mathbb{H}}\}\cup\{T_{a,\overline{a}}\circ\sigma_{\mathbb{H}} : a\in\mathbb{S}\big(Im(\mathbb{H})\big)\} := \mathcal{I}_1^+\circ\sigma_{\mathbb{H}}. &(14)
\end{aligned}$$

Proof Assertions (7), (8) are given in [8, p.215] and assertions (9), (10) are immediate consequences of the previous ones. Let $a,b\in\mathbb{S}(\mathbb{H})$, we have:

$$T_{a,b}^2 = T_{a^2,b^2},\quad T_{a,b}\circ\sigma_{\mathbb{H}} = \sigma_{\mathbb{H}}\circ T_{\overline{b},\overline{a}}$$

and then

$$\begin{aligned}(T_{a,b} \circ \sigma_{\mathbb{H}})^2 &= (T_{a,b} \circ \sigma_{\mathbb{H}}) \circ (\sigma_{\mathbb{H}} \circ T_{\overline{b},\overline{a}}) \\ &= T_{a,b} \circ T_{\overline{b},\overline{a}} \\ &= T_{a\overline{b},\overline{a}b}.\end{aligned}$$

Now

• $T_{a,b}^2 = I_{\mathbb{H}} \Leftrightarrow b^2 = \overline{a}^2 \in \{1, -1\}$. This shows assertion (11).

• $(T_{a,b} \circ \sigma_{\mathbb{H}})^2 = I_{\mathbb{H}} \Leftrightarrow a\overline{b} = \overline{b}a \in \{1, -1\} \Leftrightarrow b = \pm a$. This shows assertion (12).

Assertions (13), (14) are deduced from (11), (12). □

Proposition 5 *Let $\Phi, \Psi : \mathbb{H} \to \mathbb{H}$ be linear isometries with $\Phi(1) = 1$. Then the following two affirmations are equivalent:*

(1) $\mathbb{H} \times \mathbb{H}_{(\Phi,\Psi)}$ satisfies $(x^2, x^2, x^2) = 0$,

(2) The pair (Φ, Ψ) satisfies the following equalities:

$$\Phi\big(\Phi(x)x\big) = \Phi(x)x \text{ for all } x \in \mathbb{H}, \tag{15}$$

$$\Phi\big(\overline{x}\Psi(x)\big) = \overline{x}\Psi(x) \text{ for all } x \in \mathbb{H}, \tag{16}$$

$$\Psi\big(\Psi(y)\overline{x} + y\Phi(x)\big) = \Psi(y)\overline{x} + y\Phi(x) \text{ for all } x, y \in \mathbb{H}. \tag{17}$$

Under these conditions Φ, Ψ are involutive.

Proof Let $\odot$ be the product in algebra $\mathbb{H} \times \mathbb{H}_{(\Phi,\Psi)}$. For every $x, y \in \mathbb{H}$ we have

$$\begin{aligned}(x, y) \odot (x, y) &= \big(\Phi(x), \psi(y)\big) * (x, y) \\ &= \big(\Phi(x)x - \overline{y}\Psi(y), \Psi(y)\overline{x} + y\Phi(x)\big)\end{aligned}$$

and

$$\begin{aligned}\big((x, y) \odot (x, y)\big) \odot (1, 0) &= \big(\Phi(x)x - \overline{y}\Psi(y), \Psi(y)\overline{x} + y\Phi(x)\big) \odot (1, 0) \\ &= \Big(\Phi\big(\Phi(x)x - \overline{y}\Psi(y)\big), \Psi\big(\Psi(y)\overline{x} + y\Phi(x)\big)\Big) * (1, 0) \\ &= \Big(\Phi\big(\Phi(x)x - \overline{y}\Psi(y)\big), \Psi\big(\Psi(y)\overline{x} + y\Phi(x)\big)\Big).\end{aligned}$$

Now, Proposition 2 shows that

$$\Phi\big(\Phi(x)x - \overline{y}\Psi(y)\big) = \Phi(x)x - \overline{y}\Psi(y), \tag{18}$$

$$\Psi\big(\Psi(y)\overline{x} + y\Phi(x)\big) = \Psi(y)\overline{x} + y\Phi(x). \tag{19}$$

Equalities (15), (16) are obtained by putting $y = 0$, $x = 0$, respectively, in (18). Equality (19) is exactly (17). This shows the implication (1) $\Rightarrow$ (2). The implication (2) $\Rightarrow$ (1) is given by Proposition 2.

Now, the equality $\Phi\big(\Phi(x + 1)(x + 1)\big) = \Phi(x + 1)(x + 1)$ $(x \in \mathbb{H})$ shows immediately that Φ is involutive. The fact that Ψ is involutive follows by putting $x = 1$ in (17). □

Remarks 1 It is easy to verify the following assertions:

(1) $(I_{\mathbb{H}}, \Psi)$ satisfies (15), (16) for all $\Psi \in \mathcal{O}$,
(2) $(I_{\mathbb{H}}, I_{\mathbb{H}})$, $(\sigma_{\mathbb{H}}, \pm I_{\mathbb{H}})$ satisfy (15), (16), (17),
(3) $(I_{\mathbb{H}}, -I_{\mathbb{H}})$, $(T_{a,\overline{a}} \circ \sigma_{\mathbb{H}}, -I_{\mathbb{H}})$ do not satisfy (17) for all norm-one $a \in \mathbb{H} \setminus \mathbb{R}$,
(4) $(T_{a,\overline{a}} \circ \sigma_{\mathbb{H}}, \pm T_{b,a})$ satisfies (16), (17) for all $a, b \in \mathbb{S}\big(Im(\mathbb{H})\big)$.
(5) $(T_{a,\overline{a}} \circ \sigma_{\mathbb{H}}, I_{\mathbb{H}})$ satisfies (15), (16), (17) for all $a \in \mathbb{S}\big(Im(\mathbb{H})\big)$.
(6) $(T_{a,\overline{a}} \circ \sigma_{\mathbb{H}}, \Psi)$ satisfies (15) for all $a \in \mathbb{S}\big(Im(\mathbb{H})\big)$.

We now give an accurate list for couples (Φ, Ψ) that satisfy (15), (16), (17):

Lemma 6 *Φ satisfies (15) if and only if it belongs to $\{I_{\mathbb{H}}\} \cup \mathcal{I}_1^-$. So $I_{\mathbb{H}}$ is the only possible proper linear isometry for Φ.*

Proof Assume now that Φ satisfies (15) with $\Phi \in \mathcal{I}_1^+ \setminus \{I_{\mathbb{H}}\}$. By Lemma 5, Φ is written $T_{a,\overline{a}}$ for some $a \in \mathbb{S}\big(Im(\mathbb{H})\big)$ and the equality (15) gives $(ax)^2 = (xa)^2$ for all $x \in \mathbb{H}$. This leads to an absurdity by putting $x = a + u$ for norm-one $u \in Im(\mathbb{H}) \cap a^{\perp}$. The first affirmation follows by Proposition 5 and Remark 1(1). The second affirmation is clear. □

Let $T(x) = x + \overline{x} \in \mathbb{R}$ be the trace of $x \in \mathbb{H}$, we have $T(xy) = T(yx)$ for all $x, y \in \mathbb{H}$ [8, p. 207]. We see that $[\overline{x}, \overline{y}] = [x, y]$ for all $x, y \in \mathbb{H}$.

We need the following preliminary results.

Lemma 7 *Let a, b be norm-one in $Im(\mathbb{H})$. Then $(I_{\mathbb{H}}, T_{a,b})$ not satisfies (17).*

Proof We have

$$\begin{aligned}(I_{\mathbb{H}}, T_{a,b}) \text{ satisfies (17) } &\Rightarrow (-yb\overline{x} + ayx)b = ayb\overline{x} + yx \text{ for all } x, y \in \mathbb{H} \\ &\Rightarrow -yb\overline{x} + ayx = -ayb\overline{x}b + yx\overline{b} \text{ for all } x, y \in \mathbb{H} \\ &\Rightarrow -y + ayb = -ayb + y \text{ for } x = b \text{ and all } y \in \mathbb{H} \\ &\Rightarrow ayb = y \text{ for all } y \in \mathbb{H} \\ &\Rightarrow a = \overline{b} \in \{1, -1\}. \text{ Absurd.}\end{aligned}$$

□

The following three preliminary results provide a list for couples, solution of (15), (16), (17) of the form $(I_{\mathbb{H}}, \Psi)$, $(\sigma_{\mathbb{H}}, \Psi)$, and $(T_{a,\overline{a}} \circ \sigma_{\mathbb{H}}, \Psi)$ with $a \in \mathbb{S}\big(Im(\mathbb{H})\big)$.

Lemma 8 *$(I_{\mathbb{H}}, \Psi)$ satisfies (15), (16), (17) if and only if $\Psi \in \mathcal{E}$ where*

$$\mathcal{E} = \{I_{\mathbb{H}}\} \cup \{-T_{a,a} \circ \sigma_{\mathbb{H}} : a \in \mathbb{S}(\mathbb{H})\}.$$

Proof The if part of our Lemma follows from Remark 1(1) and the fact that $T\big(\overline{x}(\overline{y}a)\big) = T\big((\overline{y}a)\overline{x}\big)$. Assume now, for the only if part, that $(I_{\mathbb{H}}, \Psi)$ satisfies (17) and distinguish the following two cases:

(1) If $\Psi \in \mathcal{I}^+ \setminus \{I_{\mathbb{H}}, -I_{\mathbb{H}}\}$, then Ψ can be written $T_{a,b}$ for $a, b \in \mathbb{S}\big(Im(\mathbb{H})\big)$. But this cannot occur by Lemma 7.

(2) If $\Psi \in \mathcal{I}^-$, then Ψ can be written $T_{a,b} \circ \sigma_{\mathbb{H}}$ for norm-one $a, b \in \mathbb{H}$ with $b = \pm a$. The equality (17) gives

$$x\overline{b}y\overline{a}b + \overline{x}\,\overline{y}b = \overline{y}b\overline{x} + \overline{a}yx \quad \text{for all } x, y \in \mathbb{H}. \tag{20}$$

We distinguish the following two subcases:
Subcase $b = -a$. The equality (20) expresses the fact that $x(\overline{a}y)$ and $(\overline{a}y)x$ have the same trace. So (17) is true for all $x, y \in \mathbb{H}$ and

$$\psi \in \{-T_{a,a} \circ \sigma_{\mathbb{H}} : a \in \mathbb{S}(\mathbb{H})\}.$$

Subcase $b = a$. We put $y = 1$ in (20) and we have $[x, \overline{a}] = [a, \overline{x}]$.
On the other hand $[a, \overline{x}] = [\overline{a}, x] = -[x, \overline{a}]$. So $[x, \overline{a}] = 0$ for all $x \in \mathbb{H}$ and then $b = a \in \mathbb{R}$. Substituting in (20) we get $[x, y] = [\overline{y}, \overline{x}]$ for all $x, y \in \mathbb{H}$, absurd.

The result follows by Remark 1 (2), (3). □

Lemma 9 *$(\sigma_{\mathbb{H}}, \Psi)$ satisfies (15), (16), (17) if and only if $\Psi = \pm I_{\mathbb{H}}$.*

Proof The if part follows from Remark 1(2). Conversely, we show that $(\sigma_{\mathbb{H}}, \Psi)$ does not satisfy (15), (16), (17) in case $\Psi \in (\mathcal{I}^+ \setminus \{\pm I_{\mathbb{H}}\}) \cup \mathcal{I}^-$. We distinguish the following two cases:

(1) If $\Psi \in \mathcal{I}^+ \setminus \{\pm I_{\mathbb{H}}\}$) then Ψ can be written $T_{a,b}$ with $a^2 = b^2 = -1$. The equality (16) gives

$$\overline{x}axb \in \mathbb{R} \quad \text{for all } x \in \mathbb{H}. \tag{21}$$

By putting $x = 1$ in (21) we get $b = \pm a$ and therefore $(xa)^2 \in \mathbb{R}$ for all $x \in Im(\mathbb{H})$. But this cannot occur for $x = a + u$ with $u \in \mathbb{S}\big(Im(\mathbb{H})\big) \cap a^{\perp}$.

(2) If $\Psi \in \mathcal{I}^-$ then Ψ can be written $\Psi = T_{a,b} \circ \sigma_{\mathbb{H}}$ for $a, b \in \mathbb{S}(\mathbb{H})$ with $b = \pm a$. The equality (16) gives $(xa)^2 \in \mathbb{R}$ for all $x \in \mathbb{H}$, absurd.

□

Lemma 10 *Let a be a norm-one element in $Im(\mathbb{H})$. Then $(T_{a,\overline{a}} \circ \sigma_{\mathbb{H}}, \Psi)$ satisfies (15), (16), (17) if and only if $\Psi \in \mathcal{E}$ where*

$$\mathcal{E} = \{I_{\mathbb{H}}\} \cup \{\pm T_{b,a} \circ \sigma_{\mathbb{H}} : b \in \mathbb{S}\big(Im(\mathbb{H})\big)\}.$$

Proof The if part follows from Remark 1 (4), (5), (6). Now we assume that $(T_{a,\overline{a}} \circ \sigma_{\mathbb{H}}, \Psi)$ satisfies (15), (16), (17). We will distinguish the following two cases:

(1) If $\Psi = T_{b,c}$ for $b, c \in \mathbb{S}\big(Im(\mathbb{H})\big)$, then the equality (17) gives

$$by(a\overline{x}.\overline{a} - c\overline{x}.\overline{c})c = y(a\overline{x}.\overline{a} - c\overline{x}.\overline{c}) \text{ for all } x, y \in \mathbb{H}. \tag{22}$$

We distinguish the following two subcases:

(a) If $a\overline{x}.\overline{a} - c\overline{x}.\overline{c} = 0$ for all $x \in \mathbb{H}$, then $c = \pm a$

(b) If $az\overline{a} - cz\overline{c} := u \neq 0$ for some $z \in \mathbb{H}$, we put $(aza - czc)^{-1} := y$ in (22) and we get $c = -b$. For $x = z$ equality (22) gives $byu = yub$ for all $y \in \mathbb{H}$. We deduce that $bz = zb$ for all $z \in \mathbb{H}$, but this cannot occur for $b \in \mathbb{S}\big(Im(\mathbb{H})\big)$.

(2) If $\Psi = \pm T_{b,b} \circ \sigma_{\mathbb{H}}$, with $b \in \mathbb{S}(\mathbb{H})$, then (16) gives $a(\overline{b}x)^2 = (\overline{x}b)^2 a$ for all $x \in \mathbb{H}$. But this does not occur for $x = ba^{\frac{1}{2}}$ where $a^{\frac{1}{2}}$ is a square root of a.

The result follows from Remark 1 (3), (4), (5), (6). □

We can state the following.

Proposition 6 *Let (Φ, Ψ) be a pair of linear isometries of the Euclidean space $\mathbb{H}$ with $\Phi(1) = 1$. Then the algebra $\mathbb{H} \times \mathbb{H}_{(\Phi,\Psi)}$ satisfies $(x^2, x^2, x^2) = 0$ iff (Φ, Ψ) belongs to the set $\mathcal{S}_1 \cup \mathcal{S}_2 \cup \mathcal{S}_3 \cup \mathcal{S}_4 \cup \mathcal{S}_5$ where*

$$\begin{aligned}
\mathcal{S}_1 &= \{(I_{\mathbb{H}}, I_{\mathbb{H}})\},\\
\mathcal{S}_2 &= \{I_{\mathbb{H}}\} \times \{-T_{a,a} \circ \sigma_{\mathbb{H}} : a \in \mathbb{S}(\mathbb{H})\},\\
\mathcal{S}_3 &= \{\sigma_{\mathbb{H}}\} \times \{I_{\mathbb{H}}, -I_{\mathbb{H}}\},\\
\mathcal{S}_4 &= \{(T_{a,\overline{a}} \circ \sigma_{\mathbb{H}}, \pm T_{b,a}) : a^2 = b^2 = -1\},\\
\mathcal{S}_5 &= \{(T_{a,\overline{a}} \circ \sigma_{\mathbb{H}}, I_{\mathbb{H}}) : a^2 = -1\}.
\end{aligned}$$

Proof The sets $\mathcal{S}_1$, $\mathcal{S}_2$ are given by Lemma 8 and the set $\mathcal{S}_3$ is given by Lemma 9. $\mathcal{S}_4$, $\mathcal{S}_5$ are given by Lemma 10. □

6 Classification

Theorem 1 *Let $\mathcal{A}$ be an AVA with left-unit. Then the following statements are equivalent:*

(1) $\mathcal{A}$ satisfies $(x^2, x^2, x^2) = 0$,
(2) $\mathcal{A}$ satisfies $(x^2, x^2, x) = 0$,
(3) $\mathcal{A}$ is finite-dimensional and isomorphic to either $\mathbb{R}$, $\mathbb{C}$, ${}^*\mathbb{C}$, $\mathbb{H}$, ${}^*\mathbb{H}$, ${}^*\mathbb{H}(i, 1)$, $\mathbb{O}$, ${}^*\mathbb{O}$, ${}^*\mathbb{O}(i, 1)$ or $\widetilde{\mathbb{O}}$.

Proof Statements (1), (2) are equivalent by Proposition 2. Moreover, statement (3) implies (1) by Note 1, Propositions 1, 3 and Lemma 4. It then suffices to show that (1) implies (3). According to Proposition 1, it is enough to do it when $\mathcal{A}$ has dimension

≥ 8. Now, $\mathcal{A}$ is finite-dimensional of degree ≤ 4 by Lemma 1 and Proposition 2. It has dimension 8 and contains 4-dimensional subalgebras [14, Theorem 3]. It is obtained by duplication [2, Theorem 6.4], and has the form $\mathbb{H} \times \mathbb{H}_{(\Phi,\Psi)}$ where $\Phi, \Psi : \mathbb{H} \to \mathbb{H}$ are linear isometries with $\Phi(1) = 1$. We then distinguish the following cases according to Proposition 6:

(1) If $(\Phi, \Psi) \in \mathcal{S}_1$ then $\mathcal{A} = \mathbb{H} \times \mathbb{H}_{(I_{\mathbb{H}}, I_{\mathbb{H}})} = \mathbb{O}$.

(2) If $(\Phi, \Psi) \in \mathcal{S}_2$ then $\mathcal{A} = \mathbb{H} \times \mathbb{H}_{(I_{\mathbb{H}}, -T_{a,a} \circ \sigma_{\mathbb{H}})}$ with $a \in \mathbb{S}(\mathbb{H})$. Let $f = (0, 1) \in \mathbb{H} \times \mathbb{H}$. The fact that the mapping $x \mapsto (x, 0)$ is a monomorphism from algebra $\mathbb{H}$ into algebra $(\mathbb{H} \times \mathbb{H}, *) = \mathbb{O}$ allows us to write $af = (a, 0) * (0, 1) = (0, a)$ and we have $af \in \mathbb{S}\big(Im(\mathbb{O})\big)$. Now, for every $(x, y) \in \mathbb{H} \times \mathbb{H}$:

$$\begin{aligned}
\big(T_{af,\overline{af}} \circ \sigma_{\mathbb{O}}\big)(x, y) &= (0, a) * \overline{(x, y)} * (0, -a) \\
&= (0, a) * (\overline{x}, -y) * (0, -a) \\
&= (\overline{y}a, ax) * (0, -a) \\
&= (|a|^2 x, -a\overline{y}a) \\
&= (x, -a\overline{y}a) \\
&= (I_{\mathbb{H}}, -T_{a,a} \circ \sigma_{\mathbb{H}})(x, y).
\end{aligned}$$

So $\mathbb{H} \times \mathbb{H}_{(I_{\mathbb{H}}, -T_{a,a} \circ \sigma_{\mathbb{H}})}$ is equal to $\mathbb{O}_{T_{af,\overline{af}} \circ \sigma_{\mathbb{O}}}$ and is isomorphic to ${}^*\mathbb{O}(i, 1)$ by Proposition 3.

(3) If $(\Phi, \Psi) \in \mathcal{S}_3$ then $\mathcal{A}$ is either $\mathbb{H} \times \mathbb{H}_{(\sigma_{\mathbb{H}}, I_{\mathbb{H}})}$ or $\mathbb{H} \times \mathbb{H}_{(\sigma_{\mathbb{H}}, -I_{\mathbb{H}})}$. In the first case we have $\mathcal{A} \overset{\sim}{=} \mathbb{O}$. In the second one we have $\mathcal{A} = \mathbb{H} \times \mathbb{H}_{(\sigma_{\mathbb{H}}, -I_{\mathbb{H}})}$. Taking into account the fact that $(\sigma_{\mathbb{H}}, -I_{\mathbb{H}}) = \sigma_{\mathbb{O}}$, we have $\mathcal{A} = \mathbb{O}_{\sigma_{\mathbb{O}}} = {}^*\mathbb{O}$.

(4) If $(\Phi, \Psi) \in \mathcal{S}_4$ then $\mathcal{A} = \mathbb{H} \times \mathbb{H}_{(T_{a,\overline{a}} \circ \sigma_{\mathbb{H}}, \pm T_{b,a})}$ with $a^2 = b^2 = -1$. It is naturally isomorphic to $\mathbb{H} \times \mathbb{H}_{(\sigma_{\mathbb{H}}, I_{\mathbb{H}})} \overset{\sim}{=} \mathbb{O}$ by Lemmas 3, 4.

(5) If $(\Phi, \Psi) \in \mathcal{S}_5$ then $\mathcal{A} = \mathbb{H} \times \mathbb{H}_{(T_{a,\overline{a}} \circ \sigma_{\mathbb{H}}, I_{\mathbb{H}})}$ with $a^2 = -1$. For arbitrary $(x, y) \in \mathbb{H} \times \mathbb{H}$, we have

$$\begin{aligned}
T_{a,\overline{a}} \circ \sigma_{\mathbb{O}}(x, y) &= (a, 0) * \overline{(x, y)} * (\overline{a}, 0) \\
&= (a, 0) * (\overline{x}, -y) * (\overline{a}, 0) \\
&= (a\overline{x}, -ya) * (\overline{a}, 0) \\
&= (a\overline{x}\,\overline{a}, -yaa) \\
&= (a\overline{x}\,\overline{a}, y) \\
&= (T_{a,\overline{a}} \circ \sigma_{\mathbb{H}}, I_{\mathbb{H}})(x, y).
\end{aligned}$$

Hence $\mathbb{H} \times \mathbb{H}_{(T_{a,\overline{a}} \circ \sigma_{\mathbb{H}}, I_{\mathbb{H}})}$ is equal to $\mathbb{O}_{T_{a,\overline{a}} \circ \sigma_{\mathbb{O}}}$ and is isomorphic to ${}^*\mathbb{O}(i, 1)$ by Proposition 3.

□

Acknowledgements The authors thank the Professor Cándido Martín González for his remarks and suggestions.

References

1. Albert, A.A.: Absolute valued real algebras. Ann. Math. **48**, 495–501 (1947)
2. Calderón, A., Kaidi, A., Martín, C., Morales, A., Ramírez, M., Rochdi, A.: Finite-dimensional absolute-valued algebras. Israel J. Math. **184**, 193–220 (2011)
3. Chandid, A., Rochdi, A.: A survey on absolute valued algebras satisfying $(x^i, x^j, x^k) = 0$. Int. J. Algebra **2**, 837–852 (2008)
4. Chandid, A., Ramírez, M., Rochdi, A.: On finite-dimensional absolute valued algebras satisfying $(x^p, x^q, x^r) = 0$. Comm. Algebra **40**, 1525–1546 (2012)
5. Cuenca, J.A.: On one-sided division infinite-dimensional normed real algebras. Publ. Math. **36**, 485–488 (1992)
6. Diaby, K., Diankha, O., Fall, A., Rochdi, A.: On absolute-valued algebras satisfying $(x^2, y, x^2) = 0$. J. Alg. **585**, 484–500 (2021)
7. Diankha, O., Diouf, A., Ramírez, M.I., Rochdi, A.: Absolute-valued algebras with one-sided unit satisfying $(x^2, x^2, x^2) = 0$. Int. J. Algebra **7**(19), 935–958 (2013)
8. Hirzebruch, F., Koecher, M., Remmert, R.: Numbers. Springer (1991)
9. El-Mallah, M.L., El-Agawany, M.: Absolute valued algebras satisfying $(x^2, x^2, x^2) = 0$. Comm. Algebra **32**, 3537–3541 (2004)
10. Postnikov, M.: Leçons de Géométrie. Groupes et algèbres de Lie. Editions Mir, Moscow (1985)
11. Ramírez, M.I.: On four dimensional absolute valued algebras. In: Castellón, A., Cuenca, J.A., Fernández, A., Martín, C. (eds.) Proceedings of the International Conference on Jordan Structures (Málaga, June 1997), pp. 169–173. Málaga (1999)
12. Rochdi, A.: Eight-dimensional real absolute-valued algebras with left unit whose automorphism group is trivial. Int. J. Math. Math. Sci. **70**, 4447–4454 (2003)
13. Rodríguez Palacios, A.: One-sided division absolute valued algebras. Publ. Mat. **36**, 925–954 (1992)
14. Rodríguez Palacios, A.: Absolute-Valued Algebras of Degree Two. In: Non-associative Algebra and Its Applications, pp. 350–356. Kluwer Academic, Dordrecht/Boston/London (1994)
15. Rodríguez Palacios, A.: Absolute valued algebras, and absolute valuable Banach spaces. In: Advanced Courses of Mathematical Analysis I, pp. 99–155. World Scientific Publishing, Hackensack, NJ (2004)
16. Schafer, R.D.: An Introduction to Nonassociative Algebras. Academic Press (1966)

Mathematical Modeling of the Dynamics of Residual Malaria Transmission in Areas with Low Transmission: Application to Keur Socé (Senegal)

Moussa Kane, Karim Konate, Ousmane Sy, El Hadji Amadou Niang, Cheikh Gaye, Mamadou Saliou Diallo, Ibrahima Ndiaye, Yacine Aw, Almamy Youssouf Ly, Léontine Ndogou Bakhoum, Khady Ndiaye, Edouard Guédj Tine, Moussa Diop, Lucien Gning, Mor Absa Loum, Mamadou Bousso, Lassana Konate, Ousmane Faye, Oumar Gaye, and Jean Louis Ndiaye

Abstract Malaria remains a major public health concern in sub-Saharan Africa, even in areas progressing toward elimination. This study investigates malaria transmission dynamics in Keur Socé, a low-transmission setting in Senegal, using data collected between 2015 and 2021. Statistical analysis with centered moving averages

M. Kane (✉) · K. Konate · O. Sy · E. H. A. Niang · C. Gaye · M. S. Diallo · I. Ndiaye · Y. Aw · E. G. Tine · M. Diop · L. Konate · O. Faye · O. Gaye
Cheikh Anta Diop University, Dakar, Senegal
e-mail: moussa4.kane@ucad.edu.sn

K. Konate
e-mail: karim.konate@ucad.edu.sn

O. Sy
e-mail: syousmane7@gmail.com

E. H. A. Niang
e-mail: elhadjiamadou.niang@ucad.edu.sn

C. Gaye
e-mail: gaye80070@gmail.com

M. S. Diallo
e-mail: mamadousalioudiallo378@gmail.com

I. Ndiaye
e-mail: ndiayebro23@gmail.com

Y. Aw
e-mail: yacineaw89@gmail.com

E. G. Tine
e-mail: edouguedjtine@gmail.com

L. Konate
e-mail: lassana.konate@ucad.edu.sn

© The Author(s), under exclusive license to Springer Nature Switzerland AG 2026

C. T. Gueye et al. (eds.), *Mathematics, Computer Science, Cybersecurity and Artificial Intelligence*, Springer Proceedings in Mathematics & Statistics 536,
https://doi.org/10.1007/978-3-032-17666-0_2

(CMA) revealed a general downward trend in morbidity, while highlighting strong seasonality with recurrent peaks in the fourth quarter, coinciding with the post-rainy season. Entomological and molecular investigations showed a clear predominance of *Anopheles arabiensis* (>96%), and clustering analysis identified localized transmission hotspots, reflecting spatial heterogeneity of risk within the community. To interpret these patterns, a simplified SIRS–SI compartmental model was developed. Analysis of the control reproduction number (R_c) demonstrated that elimination ($R_c < 1$) is attainable with modest reductions in transmission parameters (α, β) or moderate increases in recovery (γ) and mosquito mortality (μ), consistent with interventions such as insecticide-treated nets ITNs, indoor residual spraying IRS, seasonal malaria chemoprevention SMC, rapid diagnosis, and effective treatment with artemisinin-based combination therapies ACTs. Numerical simulations and phase-plane analyses further emphasized the influence of seasonality and the critical need for continuous surveillance to prevent resurgence. These findings illustrate that mathematical modeling, when combined with entomological and statistical

O. Faye
e-mail: jogomaye@yahoo.fr

O. Gaye
e-mail: oumar.gaye@ucad.edu.sn

M. Kane · K. Konate
Department of Mathematics and Computer Science, Faculty of Science and Technology, Dakar, Senegal

O. Sy · O. Gaye
Malaria Research Capacity development, MARCAD, Dakar, Senegal

E. H. A. Niang · L. Konate · O. Faye
Laboratoire d'Ecologie Vectorielle et Parasitaire, LEVP UCAD, Dakar, Senegal

A. Y. Ly · L. N. Bakhoum · K. Ndiaye · L. Gning · M. A. Loum · M. Bousso · J. L. Ndiaye
University Iba Der Thiam of Thies, UIDT, Thies, Senegal
e-mail: ayoussouf.ly@univ-thies.sn

L. N. Bakhoum
e-mail: lndogou.bakhoum@univ-thies.sn

K. Ndiaye
e-mail: khady.ndiaye@aims-senegal.org

L. Gning
e-mail: lucien.gning@univ-thies.sn

M. A. Loum
e-mail: morabsa.loum@univ-thies.sn

M. Bousso
e-mail: mbousso@univ-thies.sn

J. L. Ndiaye
e-mail: jlndiaye@univ-thies.sn

approaches, offers a practical tool for guiding malaria control strategies in Keur Socé and other pre-elimination settings in sub-Saharan Africa.

Keywords Malaria · Mathematical modeling · Control reproduction number (R_c) · Pre-elimination settings · Keur Socé (Senegal) · Seasonal transmission · Centered moving average

1 Introduction

Malaria remains one of the oldest and deadliest diseases in human history, still responsible today for hundreds of thousands of deaths each year, mainly in sub-Saharan Africa. Historically associated with a significant share of global mortality, it continues to be endemic across vast tropical regions of Africa, Asia, and Latin America, hindering socio–economic development and representing a major public health challenge. Despite more than a century of scientific and medical efforts, and although international initiatives have significantly reduced incidence and mortality over the past two decades, progress remains highly uneven: some low-burden countries are moving toward elimination, while others, heavily affected, face stagnation or even resurgence of cases. Acknowledging these disparities, the World Health Organization (WHO) has made malaria elimination a top priority, emphasizing the need to adapt strategies to local contexts.

In Senegal, efforts by the National Malaria Control Program (PNLP) have led to a marked reduction in cases and deaths over the last decade. Localities such as Keur Socé illustrate this progress: from 136 confirmed cases in 2015, the number dropped to only 8 in 2021, with virtually no malaria-related deaths. These results reflect an encouraging trend toward pre-elimination. Yet, behind these apparent successes lies a concerning reality: the fragility of the system in the face of potential re-emergence. Indeed, in low-transmission areas, even a minor epidemic resurgence can be enough to compromise years of progress and return the disease to endemic levels.

This raises a central, simple yet crucial question: **How can we prevent the re-emergence of malaria in low-transmission settings such as Keur Socé?** To address this, we propose a mathematical modeling approach based on a compartmental SIRS–SI framework. This model not only identifies the most influential parameters driving transmission dynamics but also highlights the key intervention levers required to consolidate the pre-elimination state and prevent any reversal of progress.

2 State of the Art

2.1 History and Evolution of Malaria in the World

In this part, an assessment of the evolution of malaria is made by considering the most relevant stages which have contributed to a gradual change in research.

Malaria is a disease historically linked to humanity. It is evoked before the appearance of writing by all the great civilizations of the world [1, 2]. The conception of disease at the time, based on belief, already made a reciprocal relationship between stagnant water and the appearance of disease [1–3]. Hippocrates (460–377 BC) described a correlation between the severity of infections and marshy regions. He even concluded that splenomegaly was a symptom encountered in people who lived in the marshes. It will be necessary to wait for the eighteenth and the nineteenth century to see the first scientific contributions on the disease in particular with the conclusions of Sir Ronald Ross (1857–1932) who developed mathematical models of malaria which made it possible to eradicate malaria in Europe [4–9]. He received the Nobel Prize in 1902. But the first fundamental mathematical model, for the epidemiology of malaria influencing public health policy, was formulated by Kermack and McKendrick in 1927 [8–10]. Recall that McKendrick was a military doctor in the British army who served under the orders of Ronald Ross in 1901 in Sierra Leone during an antimalarial campaign. Their model is based on the principle of the law of mass action: They hypothesize that the appearance of new cases of infectious is proportional to the number of susceptible and infectious. And in the early 1950s, George Macdonald tested Ross's theory with epidemiological and entomological field data; he concludes that there is a threshold below which the disease dies out and above which the disease spreads [11]. Then in the 1970s, a large-scale malaria control project was launched in Garki, Nigeria [12]. In the planning of this project, a new innovative mathematical model of difference equations was developed [12]. Revisited versions of the classic epidemiological models of Ross and McKendrick have allowed an extension of the SIR model in which parameters from demography have been incorporated, such as fertility (cf. Leslie (1945), Usher (1969)), and mortality in both host and vector populations (cf. Teng et al. (2008), Yoshida and Hara (2007), Thieme and van den Driessche (1999)). And since then, many studies have been carried out on the mathematical modeling of malaria; in 2006, Chitnis et al. propose a model similar to Ngwa and Shu's model by considering constant immigration into the susceptible human class, neglecting the direct return of infectious humans to the susceptible class, and modeling a new force infection [13].

2.2 Evolution of Malaria Control Strategies in Senegal

This paragraph describes an overview of actions taken in the main areas of intervention against the malaria epidemic in Senegal. (Source: NMCP annual epidemiological activity reports and bulletins).

Already in the 1950s, during the era of the Global Malaria Eradication Program (GMEP), the French colonial administration used malariometric data and cartographic information to plan strategies to combat this scourge. In 1953, it created the Pest Control Service (PCS) based in Thiès and whose main mission was vector control and the destruction of breeding sites and mosquito breeding sites. With the advent of independence in 1960, the WHO launched the malaria pre-eradication project called OMS Senegal 13 defining the fight against malaria as a national concern. This project encouraged weekly chloroquine prophylaxis during the transmission seasons for children aged 0 to 14 and for pregnant women. The initiative lasted until 1969. Then Intra-Domiciliary Sprinkling (IDS) was used to strengthen the fight. However, around the 1980s, resistance to chloroquine in the parasite and resistance to insecticides in the vectors slowed down the dynamics of the fight. The situation did not improve, and the management of malaria became more and more complex until the creation in 1995 of the National Program for the Fight against Malaria (MNCP) by the government of Senegal, whose mission would be to coordinate and to develop new control policies and strategies on a national scale. The use of insecticide-treated mosquito nets (MII) has also been encouraged as part of the implementation of the Accelerated Malaria Control Plan (AMCP), created in 1999 with the support of the WHO. Although these interventions proved useful, problems of drug and insecticide resistance resurfaced over time. And around 2000 with the support of its financial partners, Senegal developed a national strategic control plan (spread over three periods 2001–2005, 2006–2011, and 2011–2015) which resulted in the implementation of the first campaign of distribution of ITNs in 2004, the introduction of Intermittent Presumptive Treatment (IPT), the poverty reduction strategy of 2006–2010, the use of rapid diagnostic tests (RDTs), the use of therapeutic combinations based on (CTA) throughout the country in 2006–2007, the creation and monitoring of sentinel sites in 2007, home management of malaria (PECADOM) in 2008, and the national campaign for the mass distribution of long-lasting impregnated mosquito nets duration of action 2008–2010. Over the past decade, the MNCP has emphasized operational research in order to maximize the impact of diagnostic, prevention, and therapeutic strategies.

2.3 The Limits of the Control Strategies Used

In this section, we highlight the shortcomings of the action plans implemented by the National Malaria Control Program (PNLP) and the State of Senegal.

Since independence, the State of Senegal and the PNLP have significantly reduced the total number of malaria cases and deaths. Their success has primarily relied on medication prevention, therapeutic strategies, vector control, and to some extent, poverty reduction. However, despite these notable advances, progress has stagnated in recent years: malaria continues to spread in certain regions of the country. The decreases in malaria morbidity and mortality actually conceal significant regional disparities. In 2019, the regions of Kolda, Tambacounda, and Kédougou, representing only 11% of the national population (1,822,372 inhabitants out of a total of 16,209,119), recorded 81% of confirmed malaria cases (286,001 out of 354,708 cases). Additionally, these regions accounted for 88% of cases among children under five (33,396 out of 37,941 cases) and 89% of cases among pregnant women (6,361 out of 7,155 cases). This demonstrates that the current control strategies have not been sufficient to eradicate malaria and have reached their limits. Therefore, more innovative techniques are necessary to combat the disease. Moreover, the large volume of field data collected cannot be effectively utilized without the application of mathematics. Data collection must be accompanied by a rigorous mathematical and statistical approach to enable optimal analysis and prediction, thereby minimizing risks and optimizing access to interventions in both time and space. Epidemiological analysis has long relied on mathematics, but its application remains underdeveloped in Senegal.

Through this article, we will implement a mathematical approach that allows for the exploration of scenarios inaccessible to experience, thereby informing public health decision-making. Our study will focus on Keur Socé, a pre-elimination zone for malaria, providing a valuable opportunity to inform the strategic and operational planning of control efforts. This approach will help identify the levels of intervention to target, particularly at the sub-national level, to accelerate progress toward the goals outlined in the national malaria control strategic plan. Our mathematical approach will model transmission dynamics and identify the most effective interventions. By combining field data with sophisticated mathematical models, we aim to optimize resource allocation and maximize the impact of malaria control strategies. This innovative method offers a new perspective and robust tools for malaria control in Senegal, marking a significant advance in the efforts to control and eliminate this disease.

2.4 Limitations of Previous Research and Positioning of Our Study

Over the past two decades, major advances have been made in the fight against malaria, notably through the widespread distribution of insecticide-treated nets (ITNs), the introduction of artemisinin-based combination therapies (ACTs), and the rollout of seasonal malaria chemoprevention (SMC). These interventions have substantially reduced incidence and mortality in several regions of Senegal. However, the scientific literature has long focused on high-transmission areas, aiming to understand endemic dynamics and assess the impact of large-scale interventions. In such studies, classical compartmental models (SIR, SEIR, etc.) have proven relevant for evaluating transmission thresholds, but they are often calibrated for high-intensity contexts where seasonal fluctuations and local resurgences are less critical.

In low-transmission and pre-elimination settings, such as Keur Socé in Senegal, the challenges are radically different. Here, malaria is no longer a permanent threat but a latent risk, liable to reappear in the form of sudden outbreaks. Previous studies present several limitations: They rarely address the issue of re-emergence and often overlook fine-scale entomological dimensions, such as molecular identification of vector species or the detection of localized hotspots through clustering approaches.

In this context, it is crucial to develop tailored approaches capable of answering a central question: **How can malaria re-emergence be prevented in low-transmission areas?** To address this, we propose a compartmental SIRS–SI model that integrates local, entomological, and statistical data, in order to identify the key levers of control. This approach aims to fill the gaps left by earlier studies and to provide an operational framework for consolidating pre-elimination achievements while anticipating conditions that may favor a resurgence of transmission.

3 Methodology

The work begins with the collection of morbidity and mortality data related to malaria from the health post in the commune of Keur Socé, as well as mosquito data obtained through entomological sampling techniques. These data are supplemented by the database of the Laboratory of Vectorial and Parasitic Ecology (LEVP) covering the period from 2013 to 2019. The collected data are then analyzed and processed. The next step involves presenting the data in a two-column table, facilitating graphical representation, and allowing the identification of the characteristics of the time series, particularly a general trend and seasonal variations. Subsequently, theoretical mathematical modeling is used to identify the mechanisms responsible for the observed evolution and dynamics. This involves constructing a mathematical model of malaria transmission based on a SIRS-SI model, where susceptible humans are infected by the parasite when bitten by infectious mosquitoes, and healthy mosquitoes become infected when they bite infectious humans. Additionally, the immunity of recovered humans is generally temporary. To validate our theory, epidemiological field data are used for numerical simulations and, potentially, for making predictions. The final step is to determine the techniques that allow the adjustment of the various parameters of the basic reproduction rate R0R_0R0 in order to control the disease.

3.1 Collection of Data

The municipality of Keur Socé is located in a rural area in the Kaolack region, in the health district of Ndoffane. The area is between longitudes 16°00′14.8″ and 16°07′13″ W and latitudes 13°51′53″ and 14°00′00″ N (see Fig. 1). It is located 230 km from Dakar in the Sudano-Sahelian region of Senegal. The ecology of the area is characterized by the alternation of a long dry season from November to June and a short rainy season from July to October. A demographic and health surveillance system (DSS) covering 74 villages has been set up by the Parasitology Mycology department of the Faculty of Medicine of Dakar.

An analysis of the consultation registers is carried out over the period from 2015 to 2021 to identify the number of cases of malaria confirmed by RDT and the places of origin of the patients, thus showing a distribution of the cases in clusters included in circular areas of 2 km from rays which represents the average flight distance of anopheles. Thus, three clusters were identified and in each of them the village presenting the environmental aspects in particular the presence of breeding sites positive to An. gambiae and favoring a residual transmission.

The map (Fig. 1) of the commune of Keur Socé highlights a spatial distribution of malaria cases, characterized by geographic clusters mainly located in Sagnanème, Keur Demba, and Keur Diouma, while the center of Keur Socé shows more moderate intensity and peripheral areas such as Keur Baka or Ndiédieng appear less affected. These foci are often situated near watercourses, suggesting a direct link between

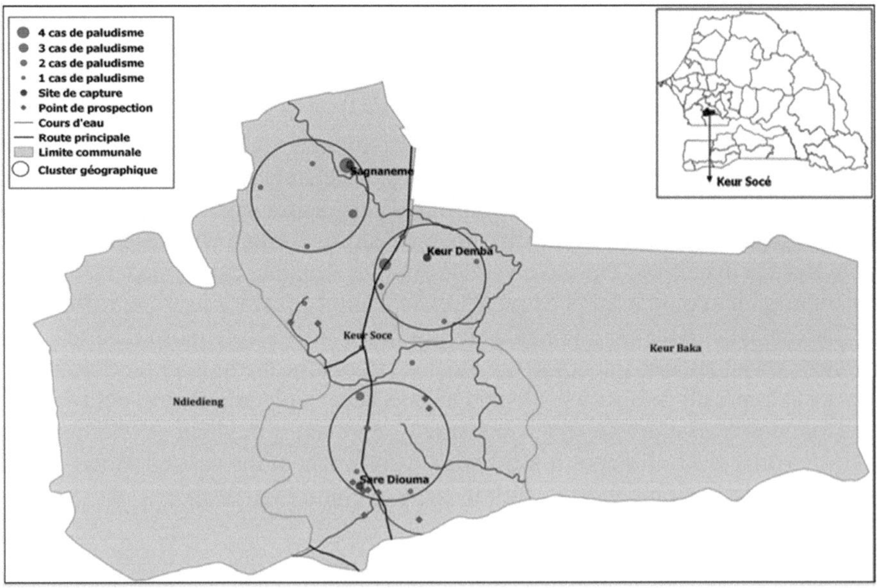

Fig. 1 Cluster mapping

transmission dynamics and the presence of larval habitats favorable to the proliferation of mosquito vectors. Capture and survey sites, strategically positioned in these areas, reflect a targeted surveillance strategy aimed at better understanding entomological dynamics and their correlation with the distribution of human cases. The identification of points with up to four cumulative cases highlights active foci which, despite an overall downward trend in malaria burden in the commune, reveal a persistent fragility, and the risk of epidemic resurgence. Thus, this cartographic representation underlines both the achievements of the pre-elimination phase and the challenges that remain, calling for differentiated and adapted interventions in high-risk areas, where the combination of epidemiological surveillance and targeted vector control appears essential to consolidate the progress made.

For entomological data, the good collaboration between the MARCAD program housed at the Faculty of Medicine and the Laboratory of Vector and Parasite Ecology (LEVP) of the Cheikh Anta Diop University in Dakar allowed the collection of data from 2013 to 2019. Data completeness was achieved through a set of activities between 2020 and 2021 funded by MARCAD. The entomological sampling techniques were carried out on the basis of nocturnal captures on humans of endophilic and/or exophilic mosquitoes and on the harvest of the residual morning fauna (Fig. 3). Collected samples were identified the following morning and dissected for infectious stages of the parasite.

Between the months of July 2021 and January 2022, a total of 2403 mosquitoes of the Anopheles genus were collected at the level of the various villages monitored, including 2297 An. Gambiae. For the moment at the laboratory level, 123 specimens including 108 An. gambiae and 15 An. funestus from the aggressive population have been identified at the molecular level (Fig. 2).

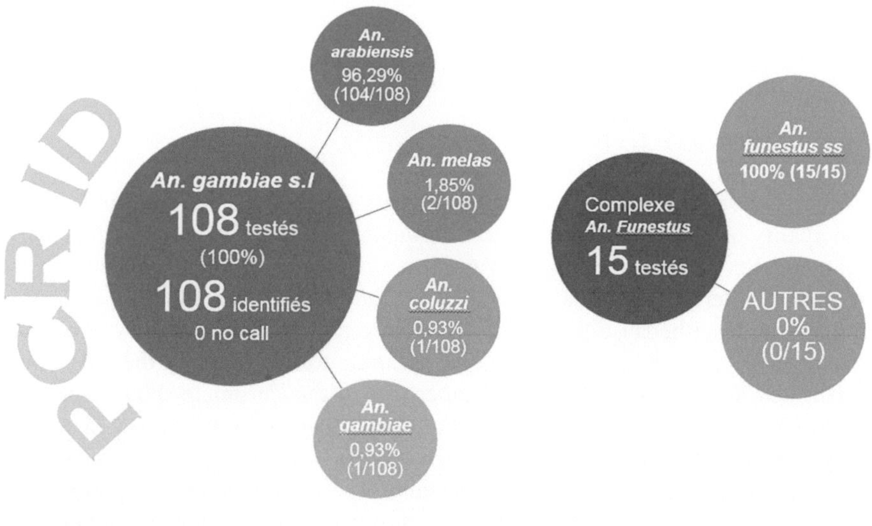

Fig. 2 PCR test results

PCR analysis showed that all *Anopheles gambiae s.l.* specimens tested (108/108) were successfully identified, with a clear predominance of *An. arabiensis* (96.29%, 104/108), followed by *An. melas* (1.85%, 2/108), *An. coluzzii* (0.93%, 1/108), and *An. gambiae s.s.* (0.93%, 1/108), all detected at very low frequencies. Within the *An. funestus* complex, all 15 individuals tested belonged exclusively to *An. funestus s.s.* (100%), with no other sibling species observed. These findings indicate a vector population largely dominated by *An. arabiensis*, alongside the exclusive presence of *An. funestus s.s.*, two major malaria vectors in the study area. Furthermore, *Plasmodium falciparum* infection screening conducted by ELISA on aggressive populations of *An. gambiae s.l.* and *An. funestus* revealed four infected mosquitoes, all belonging to *An. arabiensis*. Plasmodial infection was detected across all clusters studied, confirming the central role of this species in sustaining transmission despite an overall limited species diversity.

3.2 Mathematical Analysis by Centered Moving Average Smoothing

In this part, we discuss data analysis by smoothing centered moving averages (CMAs) in order to understand the past, analyze and explain observed values, and predict the near future.

Reminder

Let $(t_i y_i)_{1 \leq i \leq n}$ be the series of moving averages of order k. We call a series of centered moving averages of order k of the series $(t_i, y_i)_{1 \leq i \leq n}$; the series noted CMA (K) and defined by

$$CMA(k)_j = \frac{\frac{1}{2}y_{j-m} + y_{j-m+1} + \cdots + y_j + \cdots + y_{j+m-1} + \frac{1}{2}y_{j+m}}{2m}$$
$$for\, j = m + 1, \ldots, n - m. \quad (1)$$

Table 1 describes the evolution over 7 years (2015–2021) of morbidity in the municipality of Keur Socé, spread over 28 quarters. Indeed, here there is a change of time scale: We have transformed the monthly morbidity data collected at the start into quarterly data. This is justified by the fact that the monthly data were not too significant because there were many zero (0) cases of morbidity, especially for the months outside wintering. To give meaning to these values, they have been transformed into quarterly data. The scatter plot of this chronicle shows that every fourth quarter of each year there is a peak: the seasonal factor. It is repeated at equal time intervals with a period $p = 4$ (quarterly series). The seasonal factor is therefore totally determined by 4 seasonal coefficients. The amplitude of the series is not constant over time; we are therefore in the presence of a multiplicative model. The trend (in red) $y = -1.0307x + 24.944$ is determined by the method of least squares.

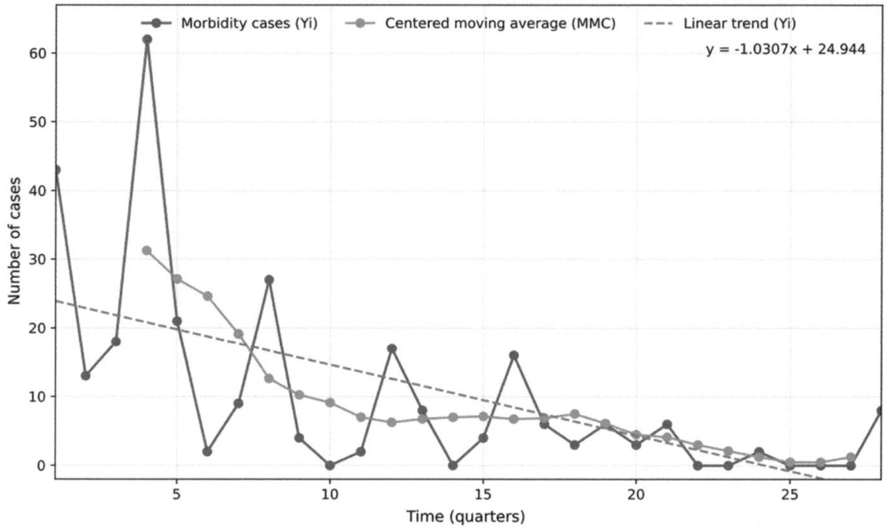

Fig. 3 Graph of morbidity and smoothing by CMA

The negative slope a = −1.0307 means that the variation in morbidity is in sharp decline. The value of the intercept 24.944 would correspond to a fictitious morbidity of about 25% in trimester 0 (Fig. 3).

Table 1 Quarterly data from 2015 to 2021

Année	Trimestre	Nb Morb Yi
	1	43
2015	2	13
	3	18
	4	62
	5	21
2016	6	2
	7	9
	8	27
	9	4
2017	10	0
	11	2
	12	17
	13	8
2018	14	0
	15	4
	16	16
	17	6
2019	18	3
	19	6
	20	3
	21	6
2020	22	0
	23	0
	24	2
	25	0
2021	26	0
	27	0
	28	8

Table 2 Trend Yi and seasonal coefficients

Année	Trimestre	Nb Morb Yi	MMC	Yi/MMC
	1	43		
2015	2	13		
	3	18	31,25	0,576
	4	62	27,125	2,28571429
	5	21	24,625	0,85279188
2016	6	2	19,125	0,10457516
	7	9	12,625	0,71287129
	8	27	10,25	2,63414634
	9	4	9,125	0,43835616
2017	10	0	7	0
	11	2	6,25	0,32
	12	17	6,75	2,51851852
	13	8	7	1,14285714
2018	14	0	7,125	0
	15	4	6,75	0,59259259
	16	16	6,875	2,32727273
	17	6	7,5	0,8
2019	18	3	6,125	0,48979592
	19	6	4,5	1,33333333
	20	3	4,125	0,72727273
	21	6	3	2
2020	22	0	2,125	0
	23	0	1,25	0
	24	2	0,5	4
	25	0	0,5	0
2021	26	0	1,25	0
	27	0		
	28	8		

Année	Trim1	Trim2	Trim3	Trim4	
2015			0,576	2,28571429	
2016	0,852791878	0,104575163	0,712871287	2,63414634	
2017	0,438356164	0	0,32	2,51851852	
2018	1,142857143	0	0,592592593	2,32727273	
2019	0,8	0,489795918	1,333333333	0,72727273	
2020	2	0	0	4	
2021	0	0			
Coeff. Saisonnier	0,872334198	0,099061847	0,883699303	3,62323115	1,36958162
Coeff.saisonniers normalisés	0,6369348	0,072330006	0,645233031	2,64550216	1

In the first quarter of each year, a 13% drop in morbidity cases is observed compared to the trend. This decrease becomes more significant every second quarter of each year, going up to 90% below the average and it continues in the third quarter with a (less considerable) decrease of 36% in morbidity cases compared to the trend. But every fourth trimester we note an increase in cases of up to more than 100% compared to the average. The model for this series is written:

$\boldsymbol{y} = (-1.0307\boldsymbol{t} + 24.944) \times \boldsymbol{CSC}$ where $\boldsymbol{CSC}$ stands for corresponding seasonal coefficient.

The purpose of this smoothing by centered moving average (CMA) is to show the shape of the trend, by removing seasonal variations. In other words, the seasonal component has been eliminated. The particularity of the analysis of a time series is precisely in the importance given to the order in which the observations are made. And the temporal dependence between the variables constitutes the main source of information thus making it possible to make predictions (Table 2).

However, it is clear that the MMC model is not satisfactory to explain the parameters that come into play in the evolution of the epidemic. In other words, it does not take into account the priority actions to be implemented to prevent and reduce cases of malaria. To overcome these shortcomings, forecasts are made using compartmentalized models.

3.3 *Formulation and Mathematical Analysis of the Compartmentalized Model*

In the following, we study a mathematical model of malaria transmission: the SIRS-SI model.

Susceptible humans become infected with the parasite when they are bitten by infectious mosquitoes and susceptible mosquitoes themselves become infected when they bite infectious humans, and furthermore the immunity of recovered humans is usually temporary. The vector component of the model does not include an immune class because mosquitoes never recover from infection, i.e., their period of infection ends with their death due to their relatively short. Thus, the immune class in the mosquito population is negligible and natural death occurs equally in all groups. Human populations are then compartmentalized into susceptible epidemiological classes S_H (t), infected $I_H(t)$, and cured or recovered $R_H(t)$. Mosquito populations are also compartmentalized into susceptible $S_m(t)$ and infected I_m (t) epidemiological classes.

The model is built on a set of assumptions mentioned as follows:

- Total human and vector population sizes are assumed to be constants.
- The development of malaria begins when the infected and infective female mosquito bites the human host.
- Individuals returned to the human population develop, as mentioned above, temporary immunity and may become susceptible again after a certain period of time.
- Susceptible mosquitoes become infected when they bite infectious humans.
- There is no vertical transmission.

The population studied is considered here as closed, i.e., no births, no deaths, no migration with spatial homogeneity. The assumption of a stable population is justified because the epidemic takes place on a small-time scale: the transmission of malaria is seasonal, it is noted only in the fourth trimester. This model therefore considers a stable or closed population. The number μ_H at the output of the susceptible cells, the infectious cells, and the cells returned to the hosts corresponds to the natural mortality rate, not linked to the disease. This choice is justified if it is considered that the disease will not have a significant effect on the overall size of the population studied.

- The number α corresponds to the force of contamination in the host, i.e., the proportion of susceptible humans who become infectious.
- The number γ designates the proportion of humans recovered among the infectious; and the immunity of recovered humans is temporary.
- Le nombre β correspond à la force de contamination chez les vecteurs, c'est-à-dire la proportion des moustiques susceptibles qui devienne infectieux.
- The number μ_m output from the susceptible and infectious cells in the vectors corresponds to the natural mortality rate, not linked to the disease.
- S_H is the population of hosts (humans) in the susceptible compartment.
- I_H is the population of hosts (humans) in the infectious compartment.
- S_m is the population of vectors (mosquitoes) in susceptible compartments.
- I_m is the population of vectors in the infectious compartments (Fig. 4).

Considering the above assumptions, the ordinary differential equations describing the dynamics of malaria in human and mosquito populations take the following form.

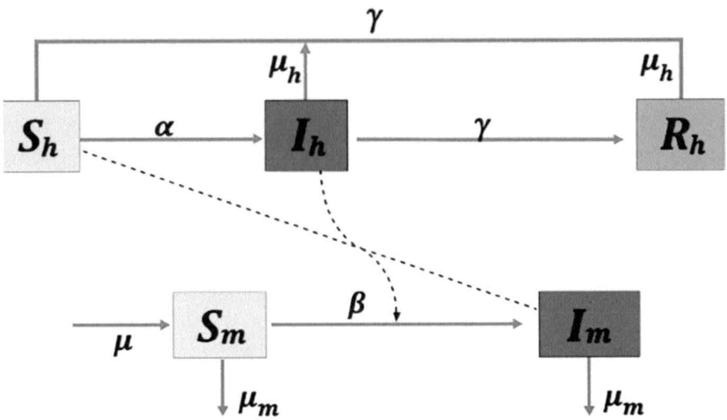

Fig. 4 Human-mosquito compartmental diagram

$$\begin{cases} \frac{dS_H}{dt} = -\alpha S_H I_m + \gamma I_H & (2.1) \\ \frac{dI_H}{dt} = \alpha S_H I_m - \gamma I_H & (2.2) \\ \frac{dS_m}{dt} = -\beta S_m I_H + \mu M - \mu S_m & (2.3) \\ \frac{dI_m}{dt} = \beta S_m I_H - \mu M & (2.4) \end{cases}$$

- $N = S_H + I_H$ is the total human population.
- $M = S_m + I_m$ is the total mosquito population.
- S_H, I_H, S_m *and* I_m are positive constants.
- $\frac{dS_H}{dt} = -\alpha S_H I_m + \gamma I_H$ here αI_m is called the force of infection. This expresses that the number of healthy individuals decreases in proportion to the number of individuals already infected and the number of individuals still uninfected.
- The proportion of infectious individuals removed from the population is γ therefore

$\frac{dI_H}{dt} = \alpha S_H I_m - \gamma I_H$ means that the number of infectious individuals' increases with those newly infected and decreases with those removed.

βS_m and μS_m are the numbers of susceptible mosquito populations suppressed per unit time due to natural death.

By adding all the differential equations, it turns out that $\frac{dS_H}{dt} + \frac{dI_h}{dt} = 0$ which means that the number of individuals in the population is always the same.

The model represented by the systems of differential equations will be analyzed in the feasible region and since the model represents populations, all state variables and parameters are assumed to be positive. The invariant region for the model is

$$\Omega_H = \left\{(S_H, I_H, R_H) \in \mathbb{R}_+^3 : S_H + I_H = N\right\} \tag{3.1}$$

$$\Omega_m = \left\{(S_m, I_m) \in \mathbb{R}_+^2 : S_m + I_m = M\right\} \tag{3.2}$$

Therefore, the solutions of the system of ordinary differential equations are feasible for all $t > 0$ if they fall into the invariant region.

$$\Omega = \Omega_H \times \Omega_m.$$

Since S_H, I_H, S_m *and* I_m are constants therefore

$$\begin{cases} \frac{dS_H}{dt} = -\alpha S_H I_m + \gamma I_H \ = 0 & (2.1) \\ \frac{dI_h}{dt} = \alpha S_H I_m - \gamma I_h \ = 0 & (2.2) \\ \frac{dS_m}{dt} = -\beta S_m I_H + \mu M - \mu S_m \ = 0 & (2.3) \\ \frac{dI_m}{dt} = \beta S_m I_H - \mu M \ = 0 & (2.4) \end{cases}$$

Then the four equations are reduced to a system of two equations

$$\begin{cases} \alpha(N - I_H)I_m - \gamma I_H = 0 & (2.5) \\ \beta(M - I_m)I_H - \mu I_m = 0 & (2.6) \end{cases}$$

This system reflects the interactions between the proportions of infected individuals and infected vectors in the host population and in the vector population respectively.

We therefore study the system in the plane $(I_H, I_m) \in [0, N] \times [0, M]$.

The expression I_m is obtained from the equation (2.5)

$$I_m = \frac{\gamma I_H}{\alpha(N - I_H)}.$$

By replacing the value of I_m in Eq. (2.6), it comes:

$$\beta\left(M - \frac{\gamma I_H}{\alpha(N - I_H)}\right)I_H - \mu\frac{\gamma I_H}{\alpha(N - I_H)} = 0.$$

Simplifying by I_H (because I_H is different from 0, in fact there is at least one infected!).

It comes:

$$\beta\left(M - \frac{\gamma I_H}{\alpha(N - I_H)}\right) - \frac{\mu\gamma}{\alpha(N - I_H)} = 0.$$

By reducing to the same denominator, it comes

$$\left[M\alpha(N - I_H) - \gamma I_H\right] - \mu\gamma = 0$$

which give

$\alpha\beta MN - \mu\gamma - \alpha\beta M I_H - \beta\gamma I_H = 0$ and $\alpha\beta MN - \mu\gamma = (\alpha\beta M + \beta\gamma)I_H.$

Hence

$$I_H = \frac{\alpha\beta MN - \mu\gamma}{\alpha\beta M + \beta\gamma}.$$

Likewise, in Eq. (2.6), it comes

$$\boxed{I_H = \frac{\mu I_m}{\beta(M - I_m)}}$$

and analogously to the previous calculation, we find

$$\boxed{I_m = \frac{\alpha\beta NM - \mu\gamma}{\alpha\beta N + \alpha\mu}}$$

- **Positivity Condition for I_H**

$$I_H > 0 \Rightarrow \alpha\beta MN - \mu\gamma > 0$$

$$\Leftrightarrow \alpha\beta MN\left(1 - \frac{\mu\gamma}{\alpha\beta MN}\right) > 0$$

$$\Longleftrightarrow 1 - \frac{\mu\gamma}{\alpha\beta MN} > 0$$

$$\Longleftrightarrow \frac{\mu\gamma}{\alpha\beta MN} < 1.$$

Let's set

$$R_0 = \frac{\alpha\beta MN}{\mu\gamma}.$$

$R_0 > 1$: then each infected individual will contaminate healthy mosquitoes after bites from the latter, and the disease will spread to an ever-increasing number of infected mosquitoes.

The $\frac{\alpha\beta MN}{\mu\gamma}$ ratio can then be interpreted as the number of contacts between susceptible mosquitoes and infectious humans capable of transmitting the disease throughout their contagion period. (Here there is contact when healthy mosquitoes bite infectious humans). Hence, by multiplying by the fraction of susceptible mosquitoes at each moment we obtain the number of new infected and infectious mosquitoes caused by a single contagious human.

- **Positivity Condition for I_m**

$$I_m > 0 \Leftrightarrow \alpha\beta MN - \mu\gamma > 0$$

$$\Longleftrightarrow \frac{\mu\gamma}{\alpha\beta MN} < 1$$

$$\Longleftrightarrow \frac{\alpha\beta MN}{\mu\gamma} > 1.$$

Let's set

$$R_0 = \frac{\alpha\beta MN}{\mu\gamma}.$$

So $R_0 > 1$ Each infected mosquito will contaminate at least one susceptible individual after bites, and the disease will spread.

The $\frac{\alpha\beta MN}{\mu\gamma}$ ratio can then be interpreted as the number of contacts between susceptible individuals and infectious mosquitoes capable of transmitting the disease throughout their contagion period (here contact is defined when susceptible humans are bitten by infectious mosquitoes).

Hence, by multiplying by the fraction of susceptible humans at each moment we obtain the number of new infected humans caused by a single contagious mosquito.

This threshold R_0 is called in the literature reproduction rate and corresponds to the average number of cases of new infected produced by an infectious host/vector during its infectious period, in a population entirely made up of susceptible hosts/vectors.

- **Local Stability of Equilibrium Points**

According to the previous system:

$$\begin{cases} \alpha(N - I_H)I_m - \gamma I_H = 0(2.5) \\ \beta(M - I_m)I_H - \mu I_m = 0.(2.6) \end{cases}$$

The Jacobean matrix is

$$J = \begin{pmatrix} -\alpha I_m - \gamma & \alpha(N - I_H) \\ \beta(M - I_m) & -\mu - \beta I_H \end{pmatrix}.$$

The trace of this matrix is

$$Tr(J) = -\alpha I_m - \gamma - \mu - \beta I_H < 0.$$

The determinant is

$$det(J) = (\alpha I_m + \gamma)(\mu + \beta I_H) - \alpha(N - I_H)\beta(M - I_m) > 0.$$

- **Stability Condition**
- **Balance Without Illness**

For $I_m = 0$ and $I_H = 0$
Trace of $(J) = -\alpha(0) - \gamma - \mu - \beta(0) < 0$

$$\Rightarrow -\gamma - \mu < 0$$

$$\iff -(\mu + \gamma) < 0$$

$$\iff \mu + \gamma > 0$$

and

$$det(J) = (\gamma)(\mu) - (\alpha N)(\beta M) > 0$$

$$\iff \gamma\mu - \alpha\beta MN > 0$$

$$\iff \gamma\mu\left(1 - \frac{\alpha\beta MN}{\gamma\mu}\right) > 0$$

$$\iff 1 - \frac{\alpha\beta MN}{\gamma\mu} > 0$$

$$\iff \frac{\alpha\beta MN}{\mu\gamma} < 1$$

$$\iff R_0 < 1 \text{ so it's stable.}$$

$R_0 < 1$ and $\det J(0, 0) > 0$ therefore both eigenvalues are positive, and the disease-free equilibrium is locally asymptotically stable. This means that introducing a small number of infected people into a large population has no chance of causing a major outbreak.

But if $R_0 > 1$ and $\det J(0, 0) < 0$ then the disease-free equilibrium is a saddle point.

- **Endemic Balance**

Trace of $(J) = -\alpha I_m - \gamma - \mu - \beta I_H < 0$

$$\iff -(\alpha I_m + \gamma + \mu + \beta I_H) < 0$$

$$\begin{aligned}\det(J) &= \mu\alpha I_m + \alpha\beta I_m I_H + \gamma\mu + \beta\gamma I_H - \alpha\beta(NM - NI_m - MI_H + I_m I_H)\\ &= \alpha\mu I_m + \beta\alpha I_m I_H + \gamma\mu + \beta\gamma I_H - \alpha\beta NM + \alpha\beta NI_m + \alpha\beta MI_H - \alpha\beta I_m I_H\\ &= -\left[\alpha\beta NM - \alpha\beta MI_H - \beta\gamma I_H - \gamma\mu\right] + \alpha\mu I_m + \alpha\beta I_m I_H + \alpha\beta NI_m - \alpha\beta I_m I_H\end{aligned}$$

But

$$\left[\alpha\beta NM - \alpha\beta MI_H - \beta\gamma I_H - \gamma\mu\right] = 0$$

$$\Rightarrow \det(J) = \alpha\mu I_m + \alpha\beta NI_m$$

$$\Longleftrightarrow det(J) > 0.$$

So

$$\alpha\mu I_m + \alpha\beta NI_m > 0$$

$$\Longleftrightarrow (\mu + \beta N)\alpha I_m > 0$$

$$\Longleftrightarrow \mu + \beta N > 0.$$

So the trace is negative and the determinant is positive so it is stable.

Hence as soon as the endemic equilibrium exists (i.e. $R_0 > 1$), it is locally asymptotically stable.

Thus, our model

$$\begin{cases}\alpha(N - I_H)I_m - \gamma I_H = 0\\ \beta(M - I_m)I_H - \mu I_m = 0\end{cases}$$

provides a mathematical framework for describing the joint dynamics of infected human and vector populations, while highlighting the central role of the basic reproduction number R_0 in determining the conditions for disease spread or extinction. To go beyond this theoretical analysis and make the dynamic mechanisms more intuitive, it is particularly relevant to represent the system graphically in the phase plane (I_H, I_m).

Such a representation, where nullclines and trajectories are displayed, makes it possible to directly visualize the possible asymptotic behaviors—disease extinction $(0, 0)$ or convergence towards an endemic equilibrium (E^*) depending on the value of R_0.

3.4 Geometry of the Nullclines (Phase-Plane)

$\dot{I}_m = 0 \Rightarrow I_m = \frac{\gamma I_H}{\alpha(N - I_H)}$: Increasing curve, vertical asymptote at $I_H \to N^-$

$\dot{I}_H = 0 \Rightarrow I_H = \frac{\mu I_m}{\beta(M - I_m)}$: Increasing, concave curve, horizontal asymptote at $I_m \to M$

Their unique intersection in the interior (if $R_0 > 1$) yields E^*.

Local epidemiological data (cases per quarter) show strong seasonality: very low transmission most of the time, but resurgences during/after the rainy season.

By representing the phase plane for parameter sets calibrated on the dry season, transition period, and rainy season, we can visualize:

- Why transmission naturally dies out in the dry season ($R_0 < 1$).
- Why it resurges at the peak of the rainy season ($R_0 > 1$).
- How interventions modify the nullclines and can bring the dynamics back below the extinction threshold even during favorable periods.

The phase-plane figures $(\boldsymbol{I_H}, \boldsymbol{I_m})$ with the two nullclines (as shown above) and trajectories for a set of parameters $(\boldsymbol{\alpha}, \boldsymbol{\beta}, \boldsymbol{\gamma}, \boldsymbol{\mu}, \boldsymbol{N}, \boldsymbol{M})$ corresponding to Keur Socé (Fig. 5).

This figure illustrates the phase plane (I_H, I_m) corresponding to the Keur Socé rainy season scenario, with a basic reproduction number $R_0 = 1.66$, which is above the epidemic threshold. The disease-free equilibrium (DFE) located at the origin (0, 0) is present but unstable, since trajectories that approach it eventually diverge away.

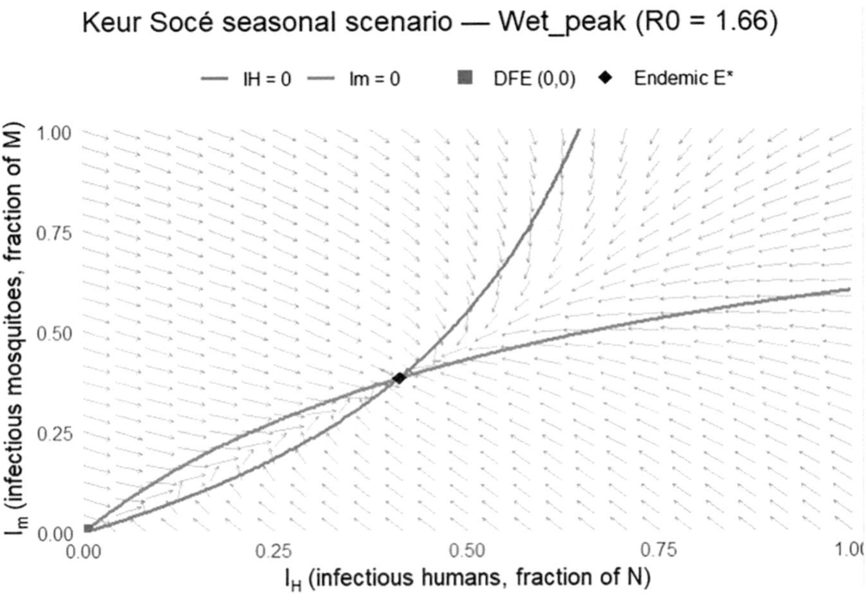

Fig. 5 Rainy season scenario, $R_0 = 1.66$

The human (blue) and vector (green) nullclines intersect at an interior point (E^*), represented by the black diamond, which constitutes the stable endemic equilibrium toward which all trajectories converge. The orange arrows of the direction field indicate that, regardless of the initial condition, the system evolves toward this endemic state characterized by non-zero proportions of infectious humans and mosquitoes. From a public health perspective, this graph highlights that during the rainy season in Keur Socé, the transmission conditions favor the persistence of malaria, and that strengthened interventions (reducing mosquito bites through bed nets and IRS, improving rapid case management, and decreasing vector longevity) are necessary to bring R_0 below 1 and make the DFE stable again (Fig. 6).

This figure shows the phase plane (I_H, I_m) for the Keur Socé transition season scenario with a basic reproduction number $R_0 = 0.71$, which is below the epidemic threshold. In this case, the disease-free equilibrium (DFE) at the origin $(0, 0)$, marked by the red square, is stable, and there is no interior endemic equilibrium. The blue and green nullclines ($\dot{I}_H = 0$ and $\dot{I}_m = 0$) do not intersect at positive values of (I_H, I_m), confirming the absence of a persistent endemic state. The direction field (orange arrows) indicates that trajectories, regardless of initial conditions, are attracted toward the origin, showing that both the fraction of infectious humans and infectious mosquitoes decline over time until extinction. From a public health perspective, this illustrates that during the transition period, transmission conditions at Keur Socé are insufficient to sustain malaria, and the system naturally tends toward

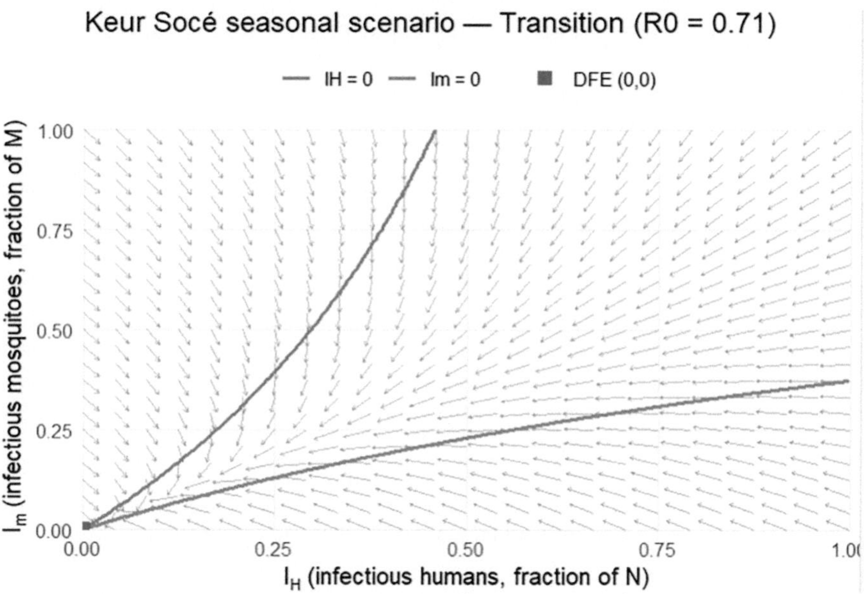

Fig. 6 Season scenario $R_0 = 0.71$

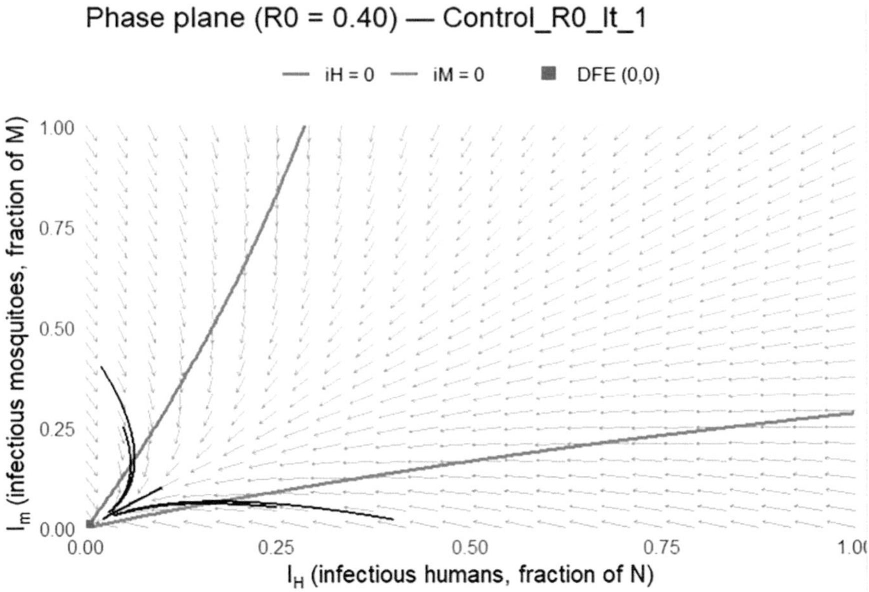

Fig. 7 Reinforced control scenario $R_0 = 0.40$

elimination, highlighting a window of opportunity where maintaining minimal interventions (bed nets, prompt diagnosis and treatment) can prevent resurgence before the rainy season (Fig. 7).

This figure illustrates the phase plane (I_H, I_m) under a reinforced control scenario with $R_0 = 0.40$, well below the epidemic threshold. In this case, the disease-free equilibrium (DFE) at the origin (0, 0), shown by the red square, is globally stable, meaning that any introduction of infectious humans or mosquitoes will eventually die out. The human (blue) and vector (green) nullclines do not intersect in the positive interior of the plane, confirming the absence of an endemic equilibrium. The direction field (orange arrows) clearly shows that trajectories, regardless of initial conditions, converge toward the origin, indicating a continuous decline in both human and mosquito infections until extinction. From a public health perspective, this graph highlights that in Keur Socé, under intensive interventions (bed nets, IRS, SMC, rapid case management), malaria transmission cannot be sustained, thereby consolidating a pre-elimination situation (Fig. 8).

This figure shows the phase plane (I_H, I_m) for the Keur Socé dry season scenario, with a basic reproduction number $R_0 = 0.23$, far below the epidemic threshold. The disease-free equilibrium (DFE) at the origin (0, 0), indicated by the red square, is globally stable, meaning that any introduction of infectious humans or mosquitoes is unsustainable and will naturally fade out. The human (blue) and vector (green) nullclines do not intersect within the positive interior of the plane, which confirms the absence of any endemic equilibrium. The direction field (orange arrows) demonstrates that trajectories from all initial conditions converge directly toward the origin,

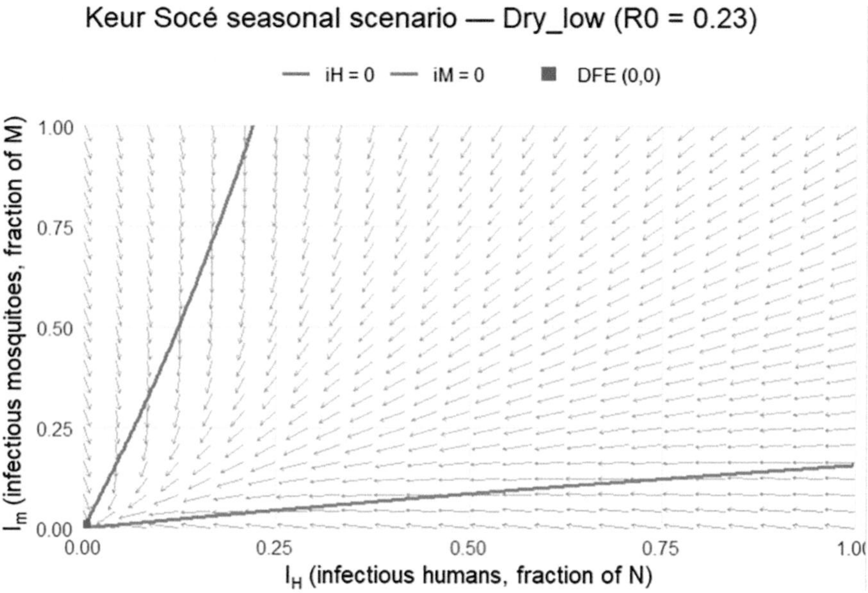

Fig. 8 Keur Socé dry season scenario $R_0 = 0.23$

reflecting a steady decline in infection levels in both populations until extinction. From a public health perspective, this graph highlights that during the dry season in Keur Socé, environmental and epidemiological conditions are insufficient to sustain malaria transmission, making this period an opportune window to consolidate control efforts and prevent reintroduction before the rains return.

After studying the geometry of the nullclines and the qualitative dynamics of the system in the phase plane, it becomes necessary to move beyond structural analysis alone in order to directly connect the mathematical model to epidemiological and operational realities. Indeed, while the nullclines allow us to visualize equilibria and the stability of endemic or disease-free states, they do not yet explain how concrete interventions can shift these equilibria. This is precisely where the analysis of the controlled reproduction number $\boldsymbol{R_c}$ becomes relevant, as it translates the effects of epidemiological parameters $(\alpha, \beta, \gamma, \mu)$, modulated by control strategies, into measurable indicators on the ground. This part is particularly useful because it establishes a direct link between the mathematical parameters of the model $(\alpha, \beta, \gamma, \mu)$ and concrete public health interventions such as the use of bed nets, IRS, rapid treatment (RDT/ACT), SMC, or larvicides, thereby quantifying their impact on the effective reproduction number $\boldsymbol{R_c}$. By expressing $\boldsymbol{R_c}$ as a function of reductions (u_α, u_β) and increases (v_γ, v_μ), it clearly identifies the priority levers and determines the conditions required to reach the elimination threshold ($\boldsymbol{R_c} < 1$). The "single-lever" thresholds provide reference points which, even if often unrealistic in isolation, make it possible to assess feasibility and guide control policies. More broadly, this approach provides a framework for optimally planning and combining multiple interventions to define strategic packages adapted to the local context, while also serving as a monitoring and evaluation tool for adjusting actions based on field data. It thus transforms the theoretical model into an operational instrument for guiding malaria control in Keur Socé and in other similar areas.

3.5 *Control Threshold R_c with Parameters Modulated by Interventions*

In this section, we transform the theoretical model into an operational tool, moving from the abstract calculation of R_0 to a practical analysis of the effect of interventions and the definition of realistic strategies for malaria elimination in Keur Socé.

Here, the part on the control threshold $\boldsymbol{R_c}$ introduces a way to integrate health interventions into the model. We start with the baseline parameters of the system (without intervention):

- $\boldsymbol{\alpha}_0$: mosquito-to-human transmission rate in the absence of control (probability that a bite infects a human).
- $\boldsymbol{\beta}_0$: human-to-mosquito transmission rate in the absence of control (probability that a mosquito becomes infectious when biting an infected human).
- $\boldsymbol{\gamma}_0$: recovery rate, spontaneous or through treatment (inverse of the average duration of human infectiousness).
- $\boldsymbol{\mu}_0$: mosquito mortality rate (inverse of their average lifespan).

Interventions (bed nets, IRS, SMC, rapid case management, etc.) modify these baseline parameters:

- $\boldsymbol{u_\alpha}$: proportionate reduction in mosquito-to-human transmission (e.g., bed nets, IRS).
- $\boldsymbol{u_\beta}$: proportionate reduction in human-to-mosquito transmission (e.g., early treatment of cases that reduces parasitemia).
- $\boldsymbol{v_\gamma}$: relative increase in the recovery rate (e.g., rapid case management, access to ACTs).
- $\boldsymbol{v_\mu}$: relative increase in vector mortality (e.g., IRS, larvicides, reduction in mosquito lifespan).

We had $R_0 = \frac{\alpha N}{\gamma} \cdot \frac{\beta M}{\mu}$. Let us call R_0^{base} the value without intervention, obtained with the baseline parameters $\alpha_0, \beta_0, \gamma_0, \mu_0$:

$$R_0^{base} = \frac{\alpha_0 N}{\gamma_0} \cdot \frac{\beta_0 M}{\mu_0}.$$

We model the interventions by:

$$\alpha = \alpha_0(1 - u_\alpha), \ \beta = \beta_0\big(1 - u_\beta\big), \ \gamma = \gamma_0\big(1 + v_\gamma\big), \ \mu = \mu_0\big(1 + v_\mu\big),$$

with $u_\alpha, u_\beta \in [0, 1]$(reductions) et $v_\gamma, v_\mu \geq 0$(increases).
The reproduction number under control is then given by

$$\begin{aligned} R_c &= \frac{\alpha N}{\gamma} \cdot \frac{\beta M}{\mu} = \frac{\alpha_0(1-u_\alpha)\mathrm{N}}{\gamma_0(1+v_\gamma)} \cdot \frac{\beta_0(1-u_\beta)M}{\mu_0(1+v_\mu)} \\ &= \frac{\alpha_0 N}{\gamma_0} \cdot \frac{\beta_0 M}{\mu_0} \cdot \frac{(1-u_\alpha)(1-u_\beta)}{(1+v_\gamma)(1+v_\mu)} \\ &= R_0^{base} \frac{(1-u_\alpha)(1-u_\beta)}{(1+v_\gamma)(1+v_\mu)}. \end{aligned}$$

The reproduction number takes the form:

$$R_c = \frac{\alpha N}{\gamma} \cdot \frac{\beta M}{\mu} = \boxed{R_0^{base} \frac{(1-u_\alpha)(1-u_\beta)}{(1+v_\gamma)(1+v_\mu)}}$$

- **Seuil d'élimination**: $R_c < 1$
- **Reduce α only ($u_\alpha > 0$, autres nuls)**

$$R_c = R_0^{base}(1-u_\alpha) < 1 \Leftrightarrow 1 - u_\alpha < \frac{1}{R_0^{base}} \Leftrightarrow \boxed{u_\alpha > 1 - \frac{1}{R_0^{base}}}$$

In a low-transmission context, where R_0^{base} is close to 1, the inequality $u_\alpha > 1 - \frac{1}{R_0^{base}}$ becomes much easier to satisfy because the threshold value is relatively low.

For example, if $R_0^{base} = 1.6$ or $R_0^{base} = 1.2$, it is sufficient to reduce α by about 37.5% or 17% respectively to achieve $R_c < 1$; and if $R_0^{base} = 1.1$, a reduction of barely 9% is enough.

This means that in low-transmission areas, even modest improvements in individual and collective protection against mosquito bites (insecticide-treated nets, IRS, repellents, environmental sanitation) can be enough to shift the dynamics below the critical threshold, making malaria elimination much more attainable.

However, this relative "ease" also implies great fragility: a relaxation of interventions or even a limited resurgence of transmission can quickly push Rc back above 1, which justifies the need to maintain strong surveillance and high coverage of control measures to consolidate progress.

- **Reduce β only**

Similarly,

$$\boxed{u_\beta > 1 - \frac{1}{R_0^{base}}}$$

This result shows that the required effort depends on the initial value of the basic reproduction number. In a low-transmission context, that is, when R_0^{base} is close to 1, the proportion of reduction needed becomes relatively modest. For example, if $R_0^{base} = 1.6$, about 37,5%. is sufficient. Concretely, this corresponds to improving the speed and effectiveness of case management in humans (early diagnosis, rapid administration of ACT, expansion of SMC) in order to reduce their infectiousness to mosquitoes. In other words, in low-transmission areas such a reduction is achievable and may be sufficient, provided it is maintained over time and combined with rigorous surveillance to prevent any resurgence.

- **Accelerating Recovery γ Alone ($v_\gamma > 0$)**

$$R_c = \frac{R_0^{base}}{(1+v_\gamma)} \langle 1 \Leftrightarrow 1 + v_\gamma \rangle R_0^{base} \Leftrightarrow \boxed{v_\gamma > R_0^{base} - 1}$$

This result means that, in a low-transmission context, a moderate improvement in the recovery rate (γ) may be sufficient to bring the dynamics below the critical threshold.

The inequality $v_\gamma > R_0^{base} - 1$ indicates that the required relative increase depends directly on the value of R_0^{base}. For example, if $R_0^{base} = 1.6$, γ must be increased by at least 0.6, i.e., a 60% rise from its baseline value, which corresponds to reducing the average infectious period by about 40%. In practice, this means that in low-transmission settings, strengthening rapid case management (widespread availability of RDTs, immediate ACT treatment, expansion of SMC) may be sufficient to interrupt the transmission chain. Thus, this result highlights that, in areas already close to elimination, improving access to and speed of treatment is a particularly effective lever to consolidate progress and prevent any resurgence.

- **Increasing Vector Mortality (μ) Alone**

  Similarly,

$$\boxed{v_\mu > R_0^{base} - 1}$$

means that, in a low-transmission context, a relatively moderate increase in mosquito mortality is sufficient to interrupt transmission.

For example, if $R_0^{base} = 1.6$, μ must increase by at least 0.6, i.e., a 60% rise in vector mortality compared to its initial level, which can be achieved through well-targeted IRS campaigns or the use of larvicides. Thus, in low-transmission areas, this result highlights that a moderate but sustained strengthening of measures aimed at reducing mosquito survival may be enough to cross the elimination threshold, provided that durable coverage is maintained and no relaxation occurs.

To conclude this section, we emphasize that the advantage in such a context is that public health efforts can be distributed across several levers, thereby avoiding reliance on a single intervention that may be difficult to sustain.

In other words, the inequality

$$\boxed{\frac{(1-u_\alpha)\big(1-u_\beta\big)}{\big(1+v_\gamma\big)\big(1+v_\mu\big)} < \frac{1}{R_0^{base}}}$$

shows that, in low-transmission settings, elimination becomes mathematically more attainable through the synergy of measures, and it provides a rational framework for planning "packages" of interventions tailored to local realities.

After having established, through theoretical analysis, the conditions required to bring the control reproduction number R_c below the critical threshold, it is now necessary to confront these results with the temporal dynamics obtained from numerical simulations. These simulations not only serve to verify the consistency of the derived inequalities but also to illustrate concretely, through scenarios calibrated on Keur Socé, the impact of different interventions on the evolution of human and vector infections. Thus, the numerical approach complements the mathematical analysis by translating elimination conditions into observable epidemiological trajectories, providing a more operational tool to guide public health decisions.

4 Numeric Simulation

The simulation is carried out by solving the SIRS model using the Euler method, implemented in Scilab. Scilab is a numerical computation software developed by INRIA, specializing in matrix manipulation, solving linear systems, and simulating differential equations, among other capabilities.

In a population of approximately 30,000 inhabitants in Keur Socé, we varied the key parameters of the model to analyze the evolution of the transmission dynamics. Particular emphasis was placed on reducing the parameter α\alpha, which represents the probability of transmission from mosquito to human. From a public health perspective, this parameter is modulated by interventions such as the consistent use of insecticide-treated nets (LLINs), indoor residual spraying (IRS), the distribution of preventive treatments (e.g., SMC in children), as well as rapid diagnosis and treatment with RDT/ACT. Mathematically, reducing $\boldsymbol{\alpha}$ results in a flattening and a delay of the epidemic peak, which operationally provides valuable time to strengthen case management, improve recovery rates, and, in the longer term, consider more radical prophylactic solutions, such as a vaccine that could directly move individuals from the susceptible compartment $\boldsymbol{S}$ to the recovered compartment RR, bypassing the infectious state $\boldsymbol{I}$.

Regarding mortality, it is not explicitly included in the core dynamics of the model, because a deceased individual, having already passed through the infectious compartment $I(t)$, no longer contributes to transmission and thus plays the same epidemiological role as a recovered individual. In the specific case of Keur

Socé, the database reports only 4 malaria-related deaths over a 7-year period (2015–2021), making their effect negligible in the overall dynamics. However, adding a $M(t)$ compartment (deaths) in exploratory simulations confirmed that reducing α also decreases the cumulative number of deaths. Therefore, acting on α\alpha not only helps control transmission but also indirectly reduces mortality, reinforcing its importance as a strategic lever in a low-transmission setting such as Keur Socé.

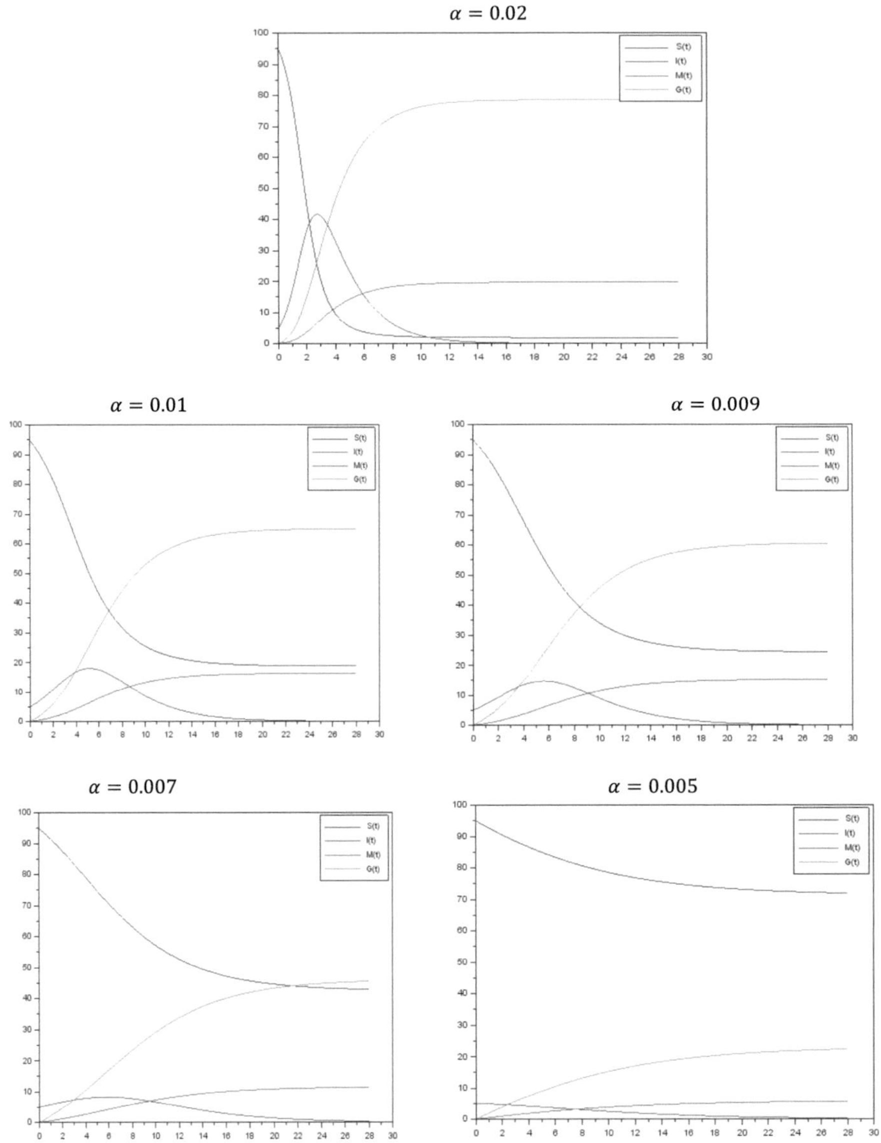

After exploring transmission dynamics using numerical simulations in Scilab, it is also relevant to compare these results with other modeling environments in order to enrich the analysis. From this perspective, the use of R software provides complementary insights, notably showing that reducing the total mosquito population (M) alone can lead to the eradication of the disease.

M= 1000

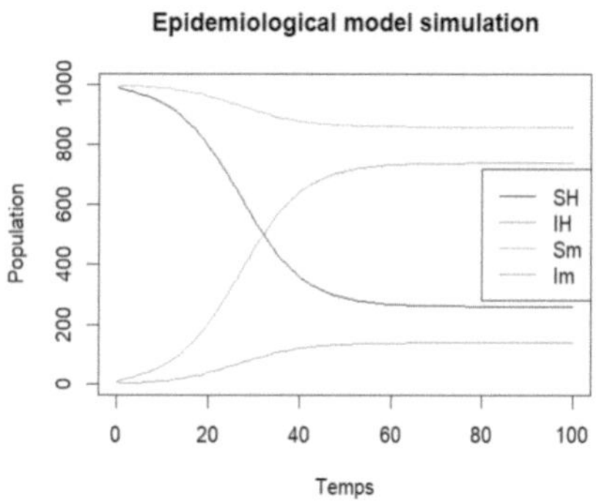

M= 800

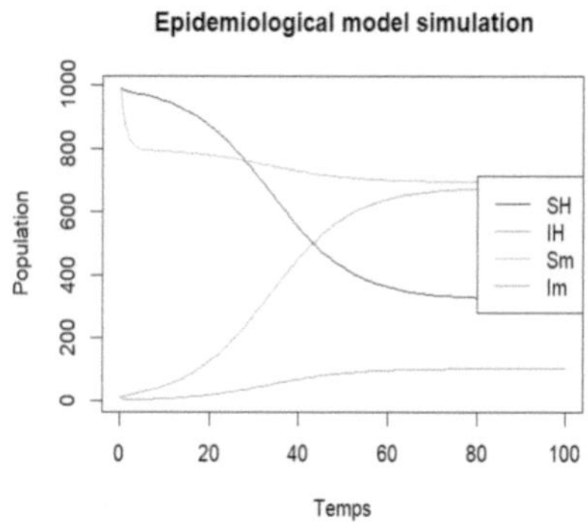

M= 500

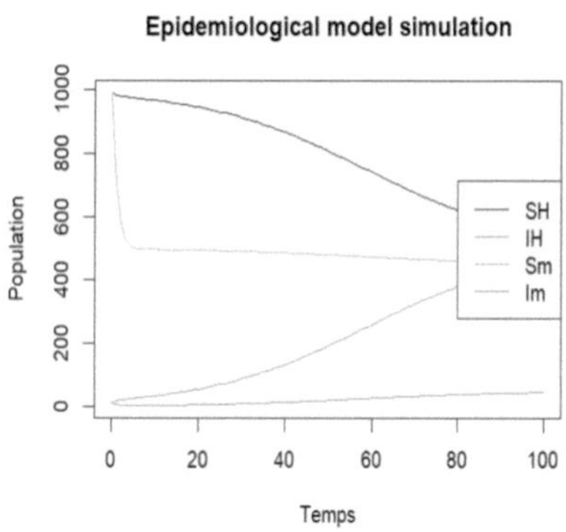

M= 300

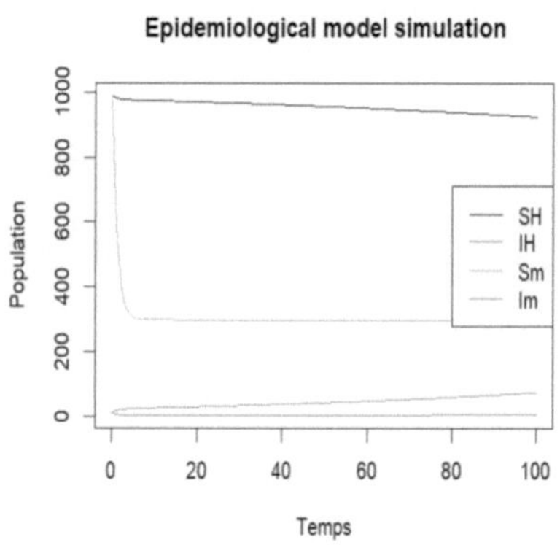

M= 200

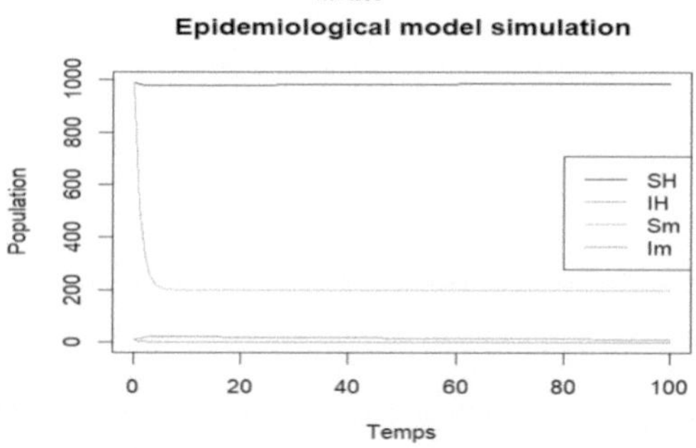

These graphs illustrate the evolution of human and vector compartments for different mosquito population sizes (M) and highlight a critical transmission threshold.

When $M = 1000$ or $M = 800$, there is a strong increase in infected humans (I_H) and mosquitoes (I_m), while susceptible humans (S_H) drop rapidly, reflecting an endemic equilibrium where the disease persists. For $M = 500$, the growth of infected individuals is slower but remains significant, confirming that transmission continues even if its intensity is reduced. In contrast, at $M = 300$, the majority of individuals remain susceptible and infections stabilize at a low level, indicating strongly limited transmission, close to extinction. Finally, for $M = 200$, the disease almost completely disappears, with S_H dominating and I_H/I_m tending toward zero. These results confirm the Ross-Macdonald model: there exists a critical threshold of vector density below which malaria dies out ($R_0 < 1$), while above this threshold the disease persists ($R_0 > 1$), thereby underlining the importance of vector control interventions to reduce M and achieve elimination.

Eradication of Malaria Through Vaccination

In our human–vector framework, the basic reproduction number is $R_0 = \frac{\alpha N}{\gamma} \cdot \frac{\beta M}{\mu}$.

If a fraction p of the human population is immunized by vaccination, the pool of susceptible individuals at the disease-free equilibrium becomes $(1-p)N$.

αN is replaced by $\alpha(1-p)N$, giving a controlled reproduction number:

$$\boxed{R_c^{vac} = (1-p)R_0}$$

The eradication threshold is then:

$$R_c^{vac} < 1 \Leftrightarrow (1-p)R_0 < 1 \Leftrightarrow \boxed{p > 1 - \frac{1}{R_0}}$$

Application to Keur Socé

- Most of the year, malaria transmission is low ($R_0 < 1$), but in the 4th quarter seasonal peak, we estimate $R_0 \simeq 1.6$.
- The vaccination threshold (with perfect vaccine efficacy) is

$$p_c = 1 - \frac{1}{1.6} \approx 0.375(\approx 37\%).$$

Thus, about 37% of the population would need to be immunized to guarantee $R_c^{vac} < 1$ during the seasonal peak. Indeed, herd immunity will be established, and an infected vector introduced into this population will transmit the disease to less than one person on average, which effectively leads to the extinction of the epidemic, as the Plasmodium parasite will encounter too many protected individuals.

5 Discussion

The fundamental objective of this study was to understand the dynamics of malaria transmission in Keur Socé and to identify the conditions necessary for its elimination. The analysis of data using centered moving averages highlighted a marked seasonal transmission pattern, with an overall downward trend over the period 2015–2021. A recurrent peak was observed in the 4th quarter of each year, indicating an intensification of transmission at the end of the rainy season. This finding confirms the critical influence of climatic factors (rainfall, humidity, temperature) already emphasized in the literature (Gupta et al., 1998; Folland et al., 1991; Ndiaye et al., 2001; Reiter, 2003; Mouchet, 2004; Alonso, 2011), as well as the role of anopheline vector ecology, whose proliferation and behavior modulate transmission risk. Consequently, vector control, particularly larval habitat management, entomological surveillance, and studies on local species (*An. gambiae, An. funestus*), remains a central component in preventing resurgence.

From a theoretical perspective, the mathematical analysis based on the basic reproduction number $R_0 = \frac{\alpha\beta MN}{\mu\gamma}$ and its controlled form R_c, allowed us to directly link the model parameters to concrete public health interventions. The results show that the elimination threshold ($R_c < 1$) can be achieved through several levers: reducing transmission parameters (α, β) via insecticide-treated nets (ITNs), IRS, or SMC; increasing the recovery rate (γ) through improved early diagnosis and rapid treatment (RDT/ACT, PECADOM); or increasing vector mortality (μ) through spraying campaigns or larvicides. Simulations indicate that in low-transmission contexts, even modest reductions ($17–37\%$ *in* α *or* β) or moderate increases ($40–60\%$ *in* γ *or* μ) are sufficient to bring dynamics below the critical threshold. However, these conditions remain fragile: relaxation of efforts or case importation can quickly reverse progress, underscoring the need for continuous surveillance and sustained intervention coverage.

Numerical simulations and phase-plane analyses confirm these analytical results. When $R_0^{base} < 1$ (dry season or under reinforced control), the disease-free equilibrium (DFE) is stable and transmission dies out. Conversely, when $R_0^{base} > 1$ (rainy season), an endemic equilibrium emerges, reflecting disease persistence. These results highlight the importance of accounting for seasonality and climate in intervention planning, as well as the necessity of adapting control strategies to seasonal peaks.

Furthermore, integrating a theoretical vaccination strategy (RTS,S/AS01) into the model underscores the potential of immunization. Vaccinating about one-third of the population in Keur Socé, where R_0 often reaches 1.5 in the 4th quarter, would be sufficient to establish herd immunity, thereby preventing the parasite from circulating sustainably. Vaccination, when combined with other control levers, could consolidate progress and accelerate the transition toward elimination.

Ultimately, this study demonstrates the operational value of mathematical modeling in malaria control. By linking theory, simulations, and local data, it provides a framework for quantifying the impact of interventions, guiding strategic priorities, and defining tailored intervention "packages." In the case of Keur Socé, our results confirm that malaria elimination is attainable, provided multifactorial, coordinated, and sustained efforts are maintained, particularly during periods of high transmission. The SIRS–SI model, by virtue of its simplicity and parsimony, offers a robust and generalizable tool to guide strategies across all low-transmission areas in Senegal and consolidate the pre-elimination state.

Study Limitations

Our model is based on a simplified compartmental structure (SIRS–SI), which facilitates analysis but does not capture the full complexity of malaria transmission, particularly asymptomatic infections, age differences, or chronic carriers. The parameters (α, β, γ, μ) were assumed constant, whereas in reality they vary with seasonality, actual intervention coverage, and vector resistance. Finally, the vaccination approach relies on a theoretical assumption of a "perfect" vaccine, while the real effectiveness of RTS,S is partial and decreases over time.

Conclusion

This study has highlighted, through the analysis of local data and the development of a simplified mathematical model (SIRS–SI), the key mechanisms that govern malaria transmission and control in Keur Socé. The results confirm that malaria dynamics are strongly shaped by seasonality, with a recurrent peak observed in the fourth quarter, directly linked to the post-rainy season. The estimation and interpretation of the control reproduction number (R_c) confirmed its role as a critical threshold: when $R_c < 1$, transmission dies out, while $R_c > 1$ sustains endemic persistence.

Numerical simulations and qualitative analyses demonstrated that, in a low-transmission context such as Keur Socé, even modest reductions in transmission parameters (α, β) or moderate increases in recovery (γ) and mosquito mortality (μ) are sufficient to push R_c below one. Moreover, the integration of vaccination strategies, even partial, significantly strengthens the feasibility of elimination through the establishment of herd immunity.

Beyond Keur Socé, the model retains substantial operational value for other pre-elimination areas with similar epidemiological characteristics. Its parsimony makes it both flexible and adaptable, allowing calibration with local data to inform health policy decisions. By quantifying the impact of interventions, ranking strategic priorities, and identifying optimal combinations of control measures, it provides a rational framework for sustaining progress towards malaria elimination.

In conclusion, the elimination of malaria in low-transmission settings is a realistic goal, provided that efforts remain combined, sustained, and adapted to seasonal variations. The proposed model proves to be an operational tool capable of guiding decision-making and strategic planning not only in Keur Socé but also in other pre-elimination foci across Senegal and sub-Saharan Africa.

Perspectives

Future research should extend this model by incorporating asymptomatic carriers, age structure, and climatic variables (rainfall, temperature, humidity) to better capture real-world malaria dynamics. Including human mobility and vector dispersion between heterogeneous transmission zones would improve the assessment of reintroduction risks.

References

1. Aurélia Stéfanie's Thesis, October 2011
2. Berge Tsanou's Thesis. Study of Some Metapopulation Epidemiological Models: Application to Tuberculosis and Malaria
3. Bichara, D.: Study of Epidemiological Models: Stability, Observation, and Parameter Estimation
4. Berhoune, H.: Master's Thesis: Mathematical Study of a Malaria Model
5. Niyukuri, F.: Mathematical Modeling of Malaria Propagation
6. Sallet, G.: Modeling and Simulation in Epidemiology INRIA & IRD 2010
7. Soufiane Bentout's Doctoral Thesis. Applied Mathematics to Some Epidemiological Models
8. Julie MINTSA MI ONDO's Thesis. Spatial Aspects in Epidemiological Modeling
9. Younsi Fatima Zohra's Thesis Presented in 2016
10. Bastin, G.: Lectures on Mathematical Modelling of Biological Systems (2012)
11. Sabatier, P., Bicout, D.J., Durand, B., Dubois, M.A.: The Use of Modeling in Animal Epidemiology
12. Jean Luc DIMI. Analysis of Epidemiological Models. Application to Intra-host Parasitic Models and the Spread of Ebola Hemorrhagic Fever
13. Pascal Zongo's Thesis. Mathematical Modeling of Malaria Transmission Dynamics

A Secure Framework for Medical Data Management Using Steganography, Cryptography, Blockchain, and AI-Based Validation

Aminata Ngom, Ndeye Fatou Ngom, and Samba Sidibe

Abstract Advances in healthcare technologies such as digital medical records, telemedicine, and artificial intelligence (AI) improve efficiency but raise cybersecurity concerns, making the protection of patient data more critical. This paper presents a secure framework that integrates steganography, cybersecurity, cryptography, and blockchain to protect the integrity and confidentiality of medical records and ensure digital trust of the transmission channel. The framework uses Stationary Wavelet Transform (SWT) and Least Significant Bit (LSB) steganography to embed compressed medical data in cover images, while RSA signatures ensure data integrity and AES encryption guarantees confidentiality. In addition, blockchain technology using smart contracts is used to immutably store the encrypted data and digital signatures, preventing tampering or unauthorized access. The system enables secure storage and retrieval of medical images and records. AI-based analysis was also performed to demonstrate that the system does not alter the intrinsic information present in the data. Clustering with k-mean and principal component analysis was used on the csv file to show that no alteration in the clustering structure. A deep convolutional network was used to predict classes of coloured images in the dataset before and after the system insertion extraction process. The results show minimal distortion in the stego images, with high PSNR and NCC values, and successful data recovery and verification through blockchain. This approach paves the way for secure medical data transmission and storage, digital sovereignty, with potential applications in telemedicine and other sensitive domains.

Keywords Cybersecurity · Blockchain · Steganography · Cryptography · Medical data security · Digital trust · Artificial intelligence

A. Ngom (✉)
LACGAA laboratory, Universitè Cheikh Anta Diop, Dakar, Senegal
e-mail: myangoma@gmail.com

N. F. Ngom · S. Sidibe
LTISI laboratory, Ecole Polytechnique de Thiès, Thies, Senegal

© The Author(s), under exclusive license to Springer Nature Switzerland AG 2026
C. T. Gueye et al. (eds.), *Mathematics, Computer Science, Cybersecurity and Artificial Intelligence*, Springer Proceedings in Mathematics & Statistics 536,
https://doi.org/10.1007/978-3-032-17666-0_3

1 Introduction

In the digital era, data security and integrity are critical, especially in healthcare, where sensitive personal, medical, and financial information is frequently targeted by cyber-attacks. Traditional security measures are often inadequate against evolving cyber threats. Modern technologies such as cryptography, blockchain, and steganography have emerged as effective solutions to enhance data protection, each providing unique methods to safeguard sensitive information [1]. This paper integrates these technologies into a comprehensive framework designed to secure medical data, focusing on embedding medical information into images through steganography, signing and encrypting the data using cryptography, and ensuring immutability through blockchain technology. Cryptography and blockchain are key technologies in healthcare data security. Cryptography secures patient information through encryption, hashing, and digital signatures, ensuring safe transmission and storage [1]. Blockchain's decentralized and immutable structure protects medical records from tampering, enhancing data integrity and traceability [2–5]. While cryptography and blockchain secure data through encryption and immutability, steganography [6] adds another layer by concealing the existence of the data. Methods such as Least Significant Bit (LSB) [7, 8], insertion Discrete Cosine Transform (DCT) [9], and Discrete Wavelet Transform (DWT) [10] have been widely employed for embedding information within digital media, preserving the media's perceptual quality while concealing the secret information [6, 11, 12]. This additional layer of security complements cryptographic techniques by adding concealment to encryption.

Recent research has explored the combination of steganography and blockchain to provide enhanced data security. For instance, OTA (*Ozyavas, Takaoglu, Ajlouni*) [3] proposes a hybrid approach combining steganography and blockchain technology. Their method divides data into fixed-sized bits and embeds them into multimedia carriers without perceptible alteration, offering enhanced resistance to steganalysis. Similarly, Pustokhina et al. [4] introduced a Blockchain-Based Secure Data Sharing Scheme (BBSDSS) for telemedicine which uses image steganography and encryption to convert medical data into stego images. These images are securely shared over a blockchain network, ensuring confidentiality and traceability of patient information. In another study, Suresh and Kamalakannan [5] proposed a method that combines blockchain with digital image steganography to provide double-layered protection for confidential data. By leveraging blockchain to transmit linked list values instead of the stego-object, their approach enhances security and mitigates vulnerabilities associated with stego-object transmission. Chaudhary et al. [11] extended these techniques by incorporating machine learning into a blockchain-based steganography framework, using algorithms such as support vector machines (SVM) and convolutional neural networks (CNN) [13] to preprocess and embed data, thus improving robustness against tampering. Raja and Suresh [14] integrates deep learning and blockchain to secure communication between IoT devices. Convolutional neural networks (CNNs) are used to hide encrypted messages in multimedia files for stealthy data transmission. The DeepStegBlock system combines this with blockchain to ensure tamper-proof and permanently recorded data exchanges. This solution caters to the security needs of IoT networks while accommodating their limited processing power, with lightweight CNNs allowing for real-time operations and enhanced data security. However, despite the progress the problem of securing medical data remains open.

This paper presents a secure framework for medical data management by integrating cryptography, steganography and blockchain technology. The framework uses Stationary Wavelet Transform (SWT) and Least Significant Bit (LSB) steganography to embed medical information in cover images. RSA cryptographic signatures are used to ensure data integrity, while AES encryption guarantees data confidentiality. The blockchain component ensures that the signed and encrypted data is immutably stored, providing an additional layer of security. AI-based analysis was also performed to demonstrate that the system does not alter the intrinsic information contained in the data. Clustering with k-mean and principal component analysis was used on the csv file to show that there was no alteration in the clustering structure. A deep convolutional network was used to predict classes of coloured images in the dataset before and after the system insertion extraction process. The proposed solution addresses key challenges in data security, particularly in telemedicine and sensitive data transmission where both confidentiality and integrity are essential.

The rest of the paper is organized as follows. Section 2 recalls the background. Section 3 presents the proposed methodology. Section 4 presents the obtained results and discusses the experiments. Section 5 concludes the work and gives some perspectives.

2 Preliminaries

This section presents the key concepts and techniques that underpin the proposed architecture for secure medical data management.

2.1 Stationary Wavelet Transform (SWT)

The Stationary Wavelet Transform (SWT) [15] is a wavelet transform designed to maintain shift invariance, addressing a limitation found in the Discrete Wavelet Transform (DWT) [10]. Unlike Discrete Wavelet Transform (DWT), which downsamples the signal during decomposition, Stationary Wavelet Transform (SWT) maintains the original resolution, making it ideal for image steganography where visual data integrity is crucial. SWT's multi-scale and multi-directional analysis enhances robustness, enhances robustness for embedding sensitive medical information while maintaining image quality, using a single-level decomposition into approximation and detail subbands The secret medical data is embedded into the approximation coefficients (cA), which represent the most visually significant parts of the image.

The SWT's shift-invariance property ensures that hidden data remains intact during image transformations, enhancing its resistance to attacks. Equation 1 governs the decomposition process.

$$
\begin{aligned}
S_{i+1}(a,b) &= \sum_j \sum_l g^i_j g^i_l S_i(a+j, b+l), \\
S^h_{i+1}(a,b) &= \sum_j \sum_l h^i_j g^i_l S_i(a+j, b+l), \\
S^v_{i+1}(a,b) &= \sum_j \sum_l g^i_j h^i_l S_i(a+j, b+l), \\
S^d_{i+1}(a,b) &= \sum_j \sum_l h^i_j h^i_l S_i(a+j, b+l),
\end{aligned}
\tag{1}
$$

where g^i_j and h^i_j are the low-pass and high-pass filters, respectively, and S_i represents the low-frequency subband at level i. The inverse SWT is used to reconstruct the original image after data embedding, preserving both image quality and embedded information.

2.2 RSA Digital Signatures

RSA [16] is an asymmetric cryptographic algorithm widely used for securing digital signatures, ensuring that data is both authentic and unaltered. In the context of our system, RSA is used to sign the *original medical data* and the *steganographed image* before they are encrypted, guaranteeing both their integrity and authenticity throughout the process. The signing process begins by creating a unique hash of the message (either the medical data or the image) using a secure hash algorithm such as SHA-256. This hash serves as a fingerprint of the original data. The hash is then encrypted using the sender's private RSA key, producing the digital signature. The signature ensures that the data has not been altered since it was signed and that it originates from a trusted source. Mathematically, the digital signature can be represented as

$$S = H(M)^d \mod n, \tag{2}$$

where M is the message, $H(M)$ is the hash of the message, d is the private key, and n is the modulus derived from RSA key generation.

To verify the signature, the recipient uses the sender's public key to compares the resulting hash with a newly calculated hash of the received message. If they match, the data's integrity is confirmed, and the sender's identity is authenticated. This process is crucial for ensuring that medical data and steganographed images remain untampered during storage or transmission. The proposed architecture applies RSA signatures to the original medical data and steganographed image before encrypting them with

AES. This allows for verification of authenticity and integrity upon decryption, as only parties with the correct public key can validate the signature, ensuring the trustworthiness of medical data throughout its lifecycle.

2.3 AES Encryption

AES (Advanced Encryption Standard) [17] is a widely used symmetric encryption algorithm, known for its strong security and efficiency, making it highly suitable for encrypting sensitive data such as medical records. In our architecture, AES plays a pivotal role in maintaining the confidentiality of both the steganographed image and the medical text data. We assume that an authentication protocol [18, 19] is already in place, which dynamically generates a unique session key for each communication session. The session key is used for encryption, providing enhanced security over traditional systems with static keys. If a session key is compromised, only that session is affected, protecting previous communications. This feature, known as forward secrecy, ensures that past sessions remain secure even if future session keys are compromised.

In the proposed framework, AES is used to encrypt the medical data, steganographed image, and their RSA signatures. This ensures that both the content and signatures remain unreadable to unauthorized parties. Only entities with the correct session key can decrypt the information, preserving data confidentiality during transmission and storage. AES (Advanced Encryption Standard) [17] is a widely used symmetric encryption algorithm, known for its strong security and efficiency, making it highly suitable for encrypting sensitive data such as medical records. In our architecture, AES plays a pivotal role in maintaining the confidentiality of both the steganographed image and the medical text data.

The AES encryption process can be summarized mathematically by Eq. 3.

$$C = E_K(P), \tag{3}$$

where P is the plaintext (the original medical data), C is the resulting ciphertext, and K is the session key. The decryption process, which restores the original data, is represented by Eq. 4.

$$P = D_K(C), \tag{4}$$

where the ciphertext C is decrypted back into plaintext using the same session key K. This encryption method ensures data security while maintaining the efficiency necessary for real-time medical applications.

2.4 Least Significant Bit (LSB) Embedding

The Least Significant Bit (LSB) embedding technique conceals secret data within an image by altering the least significant bits of pixel values, adjusting the binary representation of each pixel to embed bits of the secret message. For example, if a pixel value in binary is 11001010 and the bit to be embedded is 1, the modified value becomes 11001011. The mathematical representation of LSB embedding is given by Eq. 5.

$$\text{Pixel}_{\text{new}} = \left\lfloor \frac{\text{Pixel}_{\text{old}}}{2} \right\rfloor \times 2 + \text{bit}_{\text{secret}}. \tag{5}$$

This process is applied to the pixels of the cover image to embed the compressed medical data, ensuring that the changes remain imperceptible to the human eye.

In our architecture, LSB embedding is combined with the Stationary Wavelet Transform (SWT) to enhance the security and robustness of hidden data. The secret image is first compressed using SWT, and the approximation coefficients are then embedded into the cover image using the LSB technique. This method makes the embedded data less vulnerable to common image processing attacks like noise, compression, and cropping. The dual-layer approach ensures efficient hiding and robust recovery of sensitive data. To recover the hidden data, the system extracts the least significant bits from the pixel values of the steganographed image. The recovered bits are reassembled to retrieve the approximation coefficients (cA), followed by the application of the inverse SWT to reconstruct the original secret image. This process ensures accurate recovery of both the data and the image without loss or distortion.

2.5 Blockchain and Smart Contracts

Blockchain technology [20] provides a secure, immutable storage layer in this system, functioning as a distributed ledger that prevents data alteration or deletion. This immutability is essential for managing sensitive medical data, ensuring integrity and security. The decentralized architecture of blockchain eliminates risks linked to central points of failure and enhances traceability, allowing for secure auditing and verification of each transaction. In our system, a custom smart contract, *MedicalRecord*, is used to manage the storage of encrypted medical data and related metadata on the blockchain. Smart contracts are self-executing programs that automatically enforce the rules and logic embedded within them. The *MedicalRecord* contract is responsible for securely recording encrypted medical data, its associated hash, RSA signatures, and additional metadata, such as sensor ID and timestamp, ensuring that this information is stored immutably. By leveraging the immutable and decentralized properties of blockchain, the system ensures that all transactions are secure, transparent, and tamper-proof. This creates a robust platform for telemedicine applications, enabling secure storage and retrieval of medical data while ensuring that only authorized users can access or verify the information.

3 Methodology and Proposed Method

This section details the overall architecture of the system, explaining how cryptographic techniques, such as RSA for signing, and encryption using a session key derived from a secure authentication protocol, are combined with steganographic methods, including the Stationary Wavelet Transform (SWT) and Least Significant Bit (LSB) embedding. The system assumes that a secure authentication protocol is in place, which generates the session key used for encryption to ensure the confidentiality of the data during transmission and storage. This section also explains how secure data storage on the blockchain is achieved through smart contracts, enhancing the management of medical records in a secure and auditable environment for telemedicine. The framework integrates blockchain technology using Ganache, ensuring that data is stored immutably, meaning it cannot be altered once recorded, which is essential for maintaining the integrity of medical records. Artificial intelligence techniques, such as machine learning algorithm an deep learning model, are used to verify the framework does not alternate the intrinsnque information present in the test dataset.

3.1 System Architecture

The main stages of the architecture (Fig. 1) can be summarized into major steps:

- **Data Preparation**: Sensitive medical data, including both image and textual components, is prepared for security. The secret image is compressed using the Stationary Wavelet Transform (SWT) to reduce its size for easier embedding within a cover image. Meanwhile, textual medical data (CSV file) is loaded, signed using cryptographic methods, and prepared for encryption.
- **Steganographic Embedding**: The compressed secret image, obtained from the SWT process, is embedded into a cover image using the Least Significant Bit (LSB) method, creating a stego image that conceals the secret medical information within the cover image.
- **Signing with RSA**: The stego image, which contains the hidden secret data, is signed using RSA, ensuring the integrity of the image and allowing for later verification that the data has not been altered. Similarly, the textual medical data (CSV records) is also signed with RSA to guarantee its integrity.
- **Encryption**: Both the stego image and the signed textual data are encrypted using a session key derived from a secure authentication protocol to ensure their confidentiality during transmission and storage.

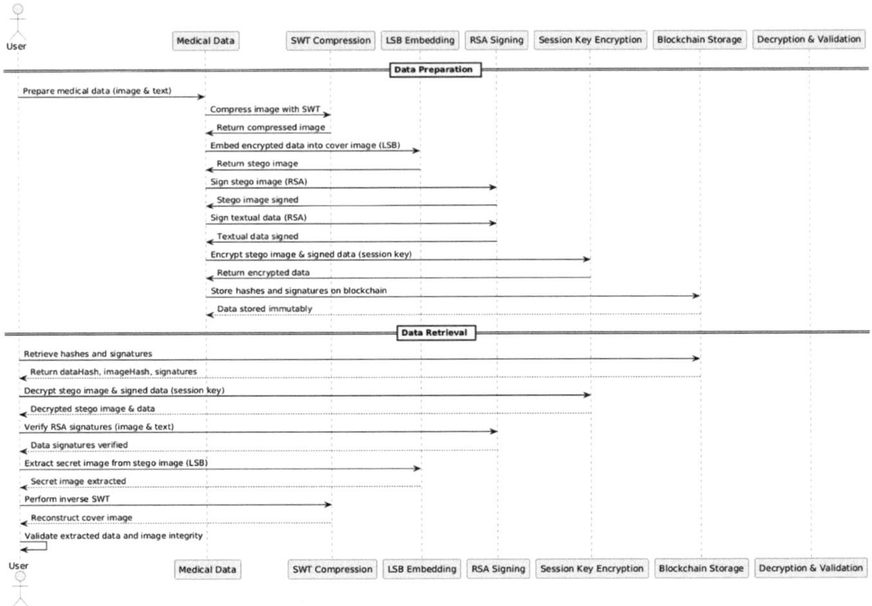

Fig. 1 Diagram of the proposed architecture

- **Blockchain Storage**: The hashes of both the stego image and the CSV data are computed and stored immutably on the blockchain, along with their corresponding RSA signatures. The blockchain-based smart contract (MedicalRecord.sol) ensures the tamper-proof storage of the medical data, making it traceable and auditable.
- **Data Retrieval and Verification**: When the data needs to be retrieved, the encrypted stego image and CSV data are fetched from the blockchain. After decryption, the RSA signatures are verified to confirm the integrity and authenticity of the data.
- **Steganographic Extraction**: Once decrypted, the secret image is extracted from the stego image using the inverse LSB technique.
- **Data Reconstruction**: Finally, the extracted secret image is decompressed using the inverse SWT process, fully reconstructing the original image for use.

3.2 Data Preparation: Signing and Encryption

In the Data Preparation (Fig. 2) signing and Encryption stage, medical data, such as patient records and sensor readings, is first structured into a format like CSV to facilitate cryptographic processing. The goal is to ensure the data's integrity and authenticity throughout its lifecycle. The process begins by generating a hash of the data using the SHA-256 algorithm. This hash acts as a unique digital fingerprint of the medical data, ensuring that any future modifications can be detected. The hash is then signed using the sender's private RSA key, which guarantees the authenticity of the data. Recipients can use the corresponding public key to verify this signature, ensuring the data has not been tampered with. To maintain confidentiality during transmission and storage, the signed data is encrypted using a session key derived from a secure authentication protocol. This session key ensures that only authorized recipients with the correct decryption key can access the sensitive medical information. The encrypted data, now both signed and secure, is ready for blockchain storage. At this point, the MedicalRecord smart contract is leveraged to store the encrypted data on the blockchain immutably. This contract is crucial for securely managing and verifying medical data and its associated information. Several elements are stored via the smart contract:

- **Data Structuring**: Medical data is formatted into a CSV-like structure to facilitate cryptographic processing.
- **Data Hashing**: A unique hash of the data is generated using the SHA-256 algorithm, serving as a digital fingerprint for integrity verification.
- **Digital Signature**: The hash is signed using the sender's private RSA key, ensuring authenticity and preventing data tampering. Verification is done using the sender's public key.
- **AES Encryption**: The signed data is encrypted using a session key derived from a secure authentication protocol, ensuring data confidentiality during storage and transmission.

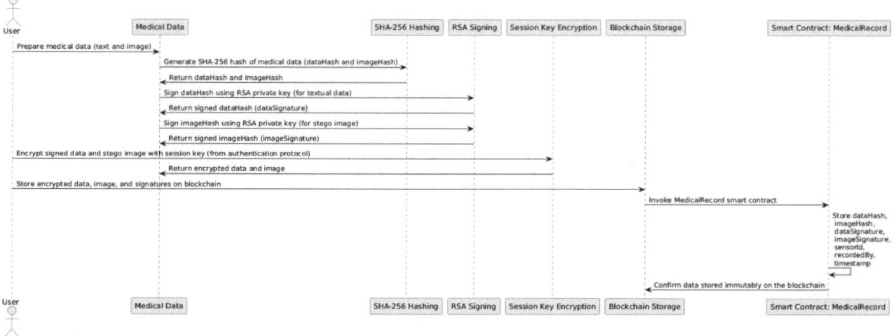

Fig. 2 Diagram for data preparation signing and encryption stage

- **Blockchain Storage**: Data is stored immutably on the blockchain using the *MedicalRecord* smart contract.

The MedicalRecord smart contract ensures that all relevant information is securely stored on the blockchain in an immutable and tamper-proof manner, preventing unauthorized alterations and providing a fully traceable record of medical data transactions.

3.3 Steganography Embedding: Concealing Data Within a Cover Image

To further enhance data security, steganography is employed to conceal the encrypted medical data within a cover image (Fig. 3). The goal of this process is to hide the existence of the sensitive information, making it challenging for potential attackers to detect that any hidden data is present. In this framework, the Stationary Wavelet Transform (SWT) is used to decompose the cover image into its approximation and detail components (Fig. ??). The encrypted medical data is embedded into the Approximation coefficients (cA), which represent the most visually significant parts of the image. This approach ensures that the quality of the cover image is preserved, and the hidden information remains imperceptible. The choice of SWT over other techniques is justified by its advantages in image processing. Unlike traditional wavelet transforms, SWT is shift-invariant, meaning it preserves spatial resolution without down-sampling the image. This makes SWT particularly suitable for image steganography, where the goal is to hide data while maintaining the quality and appearance of the cover image. Additionally, SWT captures image features in both the approximation (low-frequency) and detail (high-frequency) components, allowing for flexibility in embedding data without noticeable distortions.

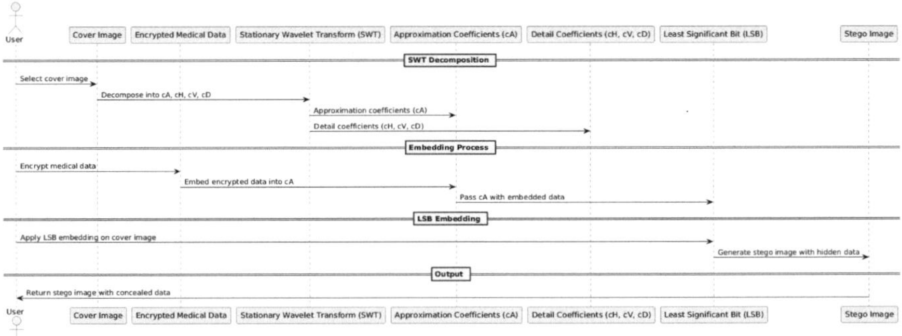

Fig. 3 Diagram of steganography embeddign

The steganographic embedding process starts with the SWT decomposition of the cover image, dividing it into four components: approximation coefficients (cA), which capture the most visually important parts, and detail coefficients (cH, cV, cD), representing horizontal, vertical, and diagonal details. The encrypted medical data is embedded into the approximation coefficients (cA) to maintain high image quality and keep the hidden data imperceptible to the human eye. After embedding the data using SWT, the Least Significant Bit (LSB) technique is applied for further concealment, modifying the least significant bits of the pixel values in the cover image to replace them with the encrypted medical data.

This approach effectively hides sensitive medical data by changing the least significant bits (LSB) of an image, which does not significantly affect its visual quality. The LSB embedding process ensures that the encrypted data is concealed while maintaining both security and image quality, as the alterations to pixel values are minimal and undetectable to the human eye. This preserves the cover image's original appearance, making the differences between the original and stego images imperceptible. By combining the strengths of the Stationary Wavelet Transform (SWT) and LSB, the system enhances data security while maintaining the cover image's integrity. SWT embeds data in the most visually significant parts, minimizing perceptual impact, while LSB further conceals the data within pixel values, ensuring sensitive information remains undetectable.

3.4 Blockchain Integration with Ganache

The integration of blockchain technology into this framework enhances the security of medical data by ensuring its immutability once stored. Ganache, a local blockchain platform, is used to simulate the Ethereum blockchain in a development and testing environment. The decentralized and immutable nature of blockchain ensures that once data is recorded, it cannot be modified or deleted, providing an additional layer of security against tampering or data loss. At the heart of this integration is a custom smart contract called *MedicalRecord*, which is designed to securely manage the storage of medical records on the blockchain. This smart contract stores several critical elements that guarantee both the authenticity and integrity of the data (Fig. 4). One key element is the *dataHash*, representing the hash of the encrypted medical data. This hash ensures the integrity of the data can be verified at any time; any modification after the data has been recorded will result in a mismatch when recalculating the hash. Similarly, an *imageHash* is generated for the steganographed image, allowing verification of the embedded image's integrity. In addition to hashing, the framework uses RSA signatures to authenticate the data. The *dataSignature* is applied to the medical data, ensuring it comes from a trusted source and has not been altered. The recipient can later use the sender's public key to verify this signature. Similarly, the *imageSignature* is used to sign the steganographed image, confirming that it has remained unchanged since it was first recorded. Another important component is the *sensorId*, a unique identifier tracing the source of the medical data to a specific

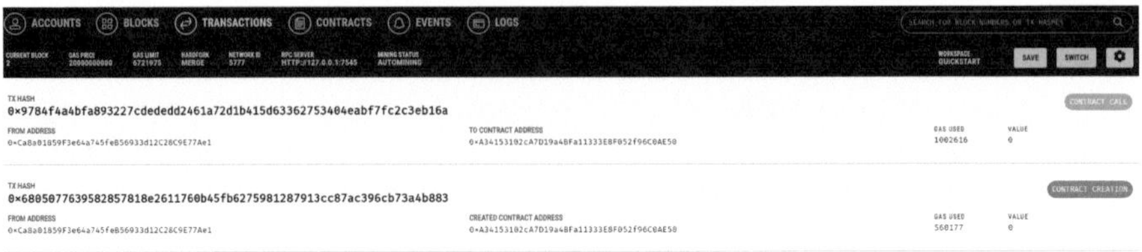

Fig. 4 Diagram of blockchain integration with ganache

device or sensor. This ensures full traceability, allowing stakeholders to confirm the origin of the data. Additionally, the *recordedBy* field stores the Ethereum address of the entity that uploaded the data to the blockchain, providing transparency and accountability. Each record is also accompanied by a *timestamp*, marking the exact time the data was stored, thereby creating a verifiable audit trail.

The first function, *storeRecord*, allows new medical data to be recorded on the blockchain. This function takes parameters such as *patientId*, *dataHash*, *imageHash*, *dataSignature*, *imageSignature*, and *sensorId*, and creates a new record. The Ethereum address of the entity recording the data is stored, and an event, *RecordStored*, is emitted to log the transaction for future reference. The second function, *getRecord*, allows users to retrieve stored medical records by providing a *patientId*. This function returns the stored data hash, image hash, RSA signatures, sensor ID, and the Ethereum address of the entity that recorded the information. It is crucial for verifying the authenticity and integrity of the stored medical records, as users can compare the retrieved hashes and signatures with the original data to confirm that nothing has been altered since its initial recording. The use of blockchain offers significant advantages in this context. First, it guarantees the immutability of stored data, meaning that medical records cannot be changed or deleted once recorded. This ensures that the integrity of the data remains intact, even in environments where malicious modifications might be attempted. Additionally, blockchain provides full traceability, as each record is associated with an Ethereum address and a timestamp, allowing stakeholders to track who performed each operation and when. Lastly, RSA signatures ensure the data's authenticity, confirming that the data originates from a trusted source and has not been tampered with. By combining advanced cryptographic techniques with blockchain's immutability, this approach offers a secure and traceable solution for managing medical records. Blockchain storage ensures that the data is protected from tampering while remaining verifiable by authorized parties, thereby ensuring a high level of security and transparency in medical data management

3.5 *Steganographic Extraction: Recovering Data from the Cover Image*

Once the encrypted and steganographed data is securely stored on the blockchain, it can be retrieved and decrypted when necessary. The process of extracting the hidden medical data and recovering the original content begins by querying the blockchain to obtain the encrypted steganographed image and its associated metadata. The system first verifies the integrity of the retrieved data by comparing the stored *dataHash* and *imageHash* with newly computed hashes from the decrypted data. This comparison ensures that the data remains unaltered during its time on the blockchain. If the hashes match, the system proceeds with the decryption process. Decryption begins by using the session key, generated from the authentication protocol, to decrypt both the steganographed image and the signed medical data. This step reverses the encryption, restoring the data to its original form and allowing for the extraction of the secret image hidden within the cover image. To ensure the data has not been tampered with, the system verifies the RSA signatures for both the medical data and the steganographed image using the sender's public key. This confirms the authenticity and integrity of the data. Once the signatures are validated, the system moves on to extract the hidden image from the steganographed cover image. The first part of this extraction involves applying the inverse Least Significant Bit (LSB) method. This technique recovers the bits of the secret image that were embedded into the cover image during the embedding phase. After retrieving the hidden bits, the system reassembles the approximation coefficients (cA) from the Stationary Wavelet Transform (SWT), which represent the most visually significant features of the image. With the approximation coefficients recovered, the inverse SWT is applied to reconstruct the original secret image. This step effectively reverses the wavelet decomposition, restoring the image to its original form with minimal quality loss. At the end of this process, both the secret image and the decrypted medical data are successfully recovered and ready for use, ensuring that sensitive information remains intact and confidential throughout its lifecycle. The secure retrieval of data from the blockchain starts with verifying its integrity. This is achieved by comparing the newly computed hashes of the retrieved data with the stored hashes. Once integrity is confirmed, the system proceeds to decrypt the medical data and steganographed image using the session key. After decryption, the RSA signatures are validated to ensure that the data has not been altered since its original recording, confirming both the authenticity and integrity of the data stored on the blockchain (Fig. 5).

3.6 *Validation with IA*

We propose to validate the proposed solution, in particular to verify that the solution does not modify, for example, the analysis results (intrinsic information) extracted from a data set (Excel file or images) containing sensitive medical information, by

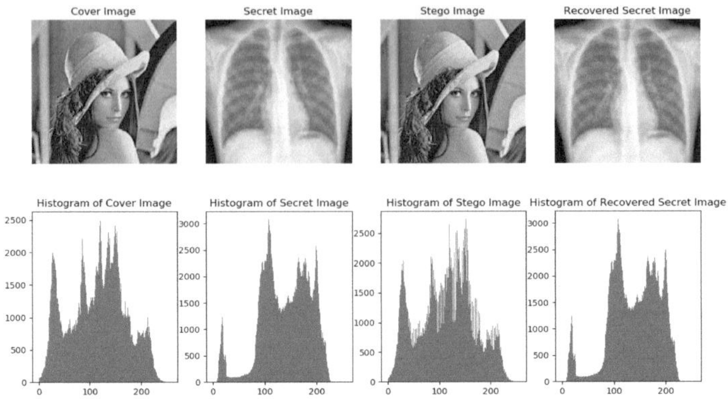

Fig. 5 Recovering data from the cover image

means of artificial intelligence models. The idea is that the doctor's or analyst's conclusion about the information hidden in the image does not change after the insertion and extraction operations on the high image. This will demonstrate that both the structure and the intrinsic information of the data are preserved by the proposed scheme. To achieve this, we propose to perform predictions using AI models before and after applying the embedding and extraction process. It will then be verified that there is no change in the conclusions of these AI models before and after the treatments.

In the case of Excel files, we propose to use the k-mean algorithm to check that the insertion and extraction operations do not change the content and shape of the clusters. To select the optimal number of clusters, the silhouette method and the elbow method are used [21]. For visual assessment of the shape of the data, the clusters will be projected into a factorial space with the minimum possible loss of information by means of a principal component analysis [21].

In the case of a dataset of images, to compare predictions before and after insertion and extraction operations, we propose to use deep neural network models [22]. Large databases of free medical images already annotated can be used to test the deep learning models for classification. Measures such as accuracy, precision, sensitivity and fscore will be used to calculate model performance [22]. In this study, we will say that there is a perfect match if the result of the AI on a test data is the same before and after extraction. Thus, if there is one perfect match for each data used to test the quality of the insertion process extraction, we will say that the match rate is 100%. This rate will be calculated on several test images with or without disease for each IA dedicated to the detection of a given disease.

3.7 Compliance with Medical Security Standards

Compliance with medical security standards like HIPAA (Health Insurance Portability and Accountability Act) [23] and GDPR (General Data Protection Regulation) [24] is essential when handling sensitive medical data. These regulations enforce strict measures to guarantee the confidentiality, integrity, and availability of patient data. For example, HIPAA mandates that data must be encrypted, with robust authentication mechanisms to prevent unauthorized access. Similarly, GDPR emphasizes data protection through encryption, data minimization, and the requirement to notify authorities in the event of a breach. Our proposed system is designed to ensure compliance with these key security objectives by incorporating RSA for digital signatures (to authenticate both medical data and images), AES for encryption, and blockchain to ensure the immutability of data. These steps collectively provide confidentiality, integrity, authentication, and non-repudiation, aligning with the requirements set forth by HIPAA and GDPR. Furthermore, session keys are used to encrypt communications, improving confidentiality and limiting the potential exposure in the case of key compromise.

In contrast, the reference system, DeepStegBlock [14], primarily focuses on AES encryption and deep learning-enhanced steganography for secure communication in IoT environments. While DeepStegBlock effectively conceals data using CNN-based steganography and ensures data integrity through blockchain, it does not incorporate digital signatures or session-based encryption to sign and verify data. Therefore, while both systems enhance data privacy and security, our solution adds an extra layer of cryptographic robustness by utilizing RSA signatures and session key-based encryption (Table 1).

4 Results and Discussion

This section provides an in-depth analysis of the results, focusing on three key metrics: PSNR, NCC, and SSIM. These results help evaluate the visual quality of the steganographed image, the fidelity of the recovered images, and the efficiency of securing data via blockchain.

4.1 Visual Quality of Steganographed Images

To measure how the data embedding process affects the visual quality of the cover image, the *Peak Signal-to-Noise Ratio* (PSNR) is used. A high PSNR value indicates that the embedding process has only a minimal effect on the image. In our tests, the PSNR value was *51.15 dB*, confirming that the cover image retained its original quality even after embedding sensitive medical data. This result highlights the effec-

Table 1 Compliance with security objectives according to HIPAA and GDPR for our system and DeepStegBlock

Security objective	Our system	DeepStegBlock [14]
Data confidentiality	AES encryption ensures both medical images and text data are protected during transmission and storage	Encrypted messages may be embedded within multimedia content
Data integrity	RSA signatures verify that medical data has not been altered during the process	Relies on blockchain's immutability to ensure that data is not altered after recording
Data availability	Blockchain ensures data availability through decentralized storage, preventing data loss or tampering	Uses blockchain to ensure the availability and immutability of the data
Encryption of sensitive data	AES encryption is used for both the steganographed image and medical data, ensuring data remains unreadable to unauthorized parties	Encrypted messages are embedded within multimedia content via steganography, though the steganographed image itself may not be encrypted
Authentication and access control	RSA signatures authenticate the sender's identity, while session keys control access	No specific mention of cryptographic authentication or access control mechanisms
Audit controls and monitoring	Blockchain provides a full audit trail with timestamps and the address of the entity storing the data	Uses blockchain to log data transactions, providing traceability and auditability
Data minimization	Only essential data (hashes, signatures) are stored on the blockchain to minimize exposure of sensitive data	No specific mention of data minimization techniques
Data breach notification	Blockchain and session-based encryption ensure unauthorized access or tampering is detectable through hash mismatch and signature verification	No specific mention of breach notification, though blockchain ensures traceability of data transactions

tiveness of using the *Stationary Wavelet Transform* (SWT) and Least Significant Bit (LSB) techniques for embedding data. The high PSNR value means that the visual distortion introduced by the embedding process is imperceptible to the human eye, ensuring that the hidden data remains undetectable without compromising the quality of the cover image.

4.2 *Fidelity of Recovered Data*

Two metrics were used to assess the accuracy of the data recovery process: *Normalized Cross-Correlation* (NCC) and the *Structural Similarity Index* (SSIM). Both metrics measure the similarity between the original secret image and the recovered image. The NCC value obtained was *0.999999*, demonstrating that the system accurately retrieves the hidden medical data. This high accuracy reflects the success of the steganographic process from embedding to recovery. The SSIM between the original and recovered secret images was *0.9999987*, confirming the high visual fidelity of the recovered images. This result shows that not only is the recovery of the covert data accurate, but also that the visual characteristics of the covert data are preserved. Furthermore, the SSIM between the cover image and the steganographed image was *0.9976*. This indicates that the embedding of the data did not significantly affect the integrity of the cover image.

4.3 *Securing Medical Data with Blockchain and Smart Contracts*

Our system uses blockchain to securely store medical data. Through the *MedicalRecord* smart contract, several critical elements are stored, including the hashes of the encrypted medical data (dataHash) and the steganographed image (imageHash), the RSA signatures, the unique sensor ID (sensorId), and the blockchain address of the entity that recorded the data. Storing data on the blockchain ensures immutability, meaning that once recorded, the data cannot be altered or deleted. Upon retrieval, the system checks the integrity of the data by comparing the stored hashes with newly calculated ones. If the hashes match, the system proceeds to decrypt the steganographic image and medical data using the session key. The RSA signatures are verified to ensure that neither the medical data nor the steganographed image have been tampered with, thus ensuring the authenticity and integrity of the retrieved data. The secret image is extracted from the steganographed cover image using the inverse LSB technique. After extracting the hidden bits, the system reconstructs the image using the approximation coefficients (cA) from the SWT. The inverse SWT is then applied to recover the original secret image with minimal quality loss. Finally, both the secret image and the decrypted medical data are successfully recovered, maintaining a high level of security and integrity.

4.4 *Processing and Blockchain Transaction Times*

In the work of DeepStegBlock [14], an innovative solution is proposed that combines deep learning-enhanced steganography with blockchain technology for secure

Table 2 Comparison between DeepStegBlock and our method

Method	DeepStegBlock	Our method
PSNR (dB)	70.3	51.148
SSIM (recovered image)	0.98	0.9999987
NCC	Not specified	0.999999
SSIM (cover image)	Not specified	0.9976
Processing time (s)	2.5	2.459
Blockchain transaction time	7 min	0.0946 s
Signature usage	No	Yes
Data type	Image	Text and image

communication in IoT devices. One of the key strengths of this approach is the use of convolutional neural networks (CNNs) to hide encrypted messages within multimedia content, ensuring undetectable transmission. In terms of visual quality, DeepStegBlock achieves a PSNR of *70.3 dB*, indicating excellent quality for steganographed images, while the SSIM between the original and steganographed images reaches *0.98*, highlighting the high fidelity of the recovered images. Our method also combines steganography with blockchain, but includes compression steps using *Stationary Wavelet Transform* (SWT) and LSB insertion to hide the secret image. While our PSNR is lower at *51.148 dB*, the SSIM and NCC metrics show better fidelity between the recovered secret images and the original ones. We obtained an SSIM of *0.9999987* and an almost perfect NCC of *0.999999*, demonstrating the precise recovery of the secret image after extraction. In terms of processing time, our methods are reasonably equal, with *2.459* s compared to *2.5* s for DeepStegBlock. However, our blockchain transaction time is significantly faster at only *0.0946* s compared to DeepStegBlock's *7* min, making our system much more efficient for real-time applications (Table 2).

4.5 IA Analyis

In order to analyse the quality of information preservation after the insertion and extraction process, two types of dataset are used. The first dataset is a csv lung cancer dataset, which contains data on lung cancer mortality [25]. The database includes demographic information, medical history, cancer diagnosis and treatment-related variables, capturing essential details about each patient's condition and history. Using k-mean clustering methods, we were able to highlight the importance of variables such as age, body mass index (BMI), cholesterol levels, family history of cancer, presence of other cancers and smoking, family history of cancer, presence of other cancers and smoking status in classifying patients. To make the clusters more visible, the data are first projected into a forward factorial space, the clustering stage.

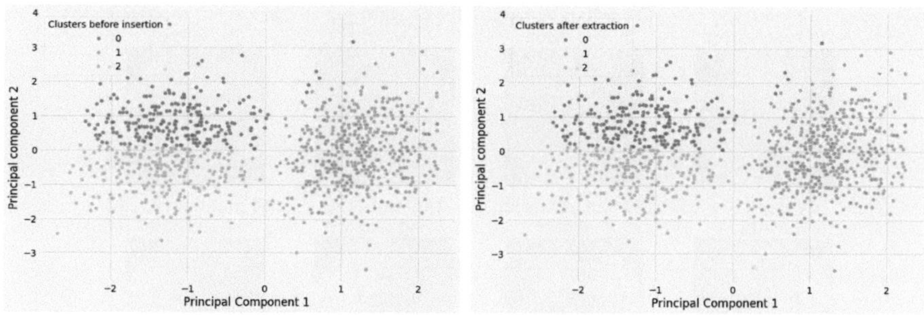

Fig. 6 Shape of clusters before and after insertion of the csv file

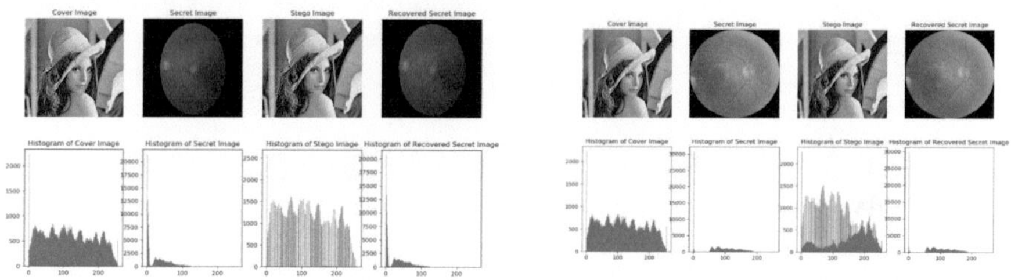

Fig. 7 Example of coloured images before insertion, inserted and after extraction with their histograms

Figure 6 shows the shape of the clusters before and after the insertion and extraction process, showing no changes in cluster composition. Cluster 0 (249 individuals) consists of older individuals with high BMI, high cholesterol and a history of other cancers, mostly passive or former smokers. Cluster 1 (492 individuals) includes middle-aged individuals with similar health profiles. Cluster 2 (259 individuals) includes younger individuals with normal BMI and cholesterol, minimal cancer history, and mainly non-smokers.

The second dataset used in this study is the Ocular Disease Intelligent Recognition (ODIR) dataset [26], which contains data from 5,000 patients. It includes patient age, colour fundus photographs of both eyes, and diagnostic keywords provided by clinicians. Patients are grouped into eight categories: Normal (N), Diabetes (D), Glaucoma (G), Cataract (C), Age-related Macular Degeneration (A), Hypertension (H), Pathological Myopia (M), and Other Diseases (O).

Figure 7 illustrates an example of an images obtained before insertion, after insertion, and after extraction. It is clear that, although there are slight changes in the histogram of the host image, there is no noticeable difference between the histograms of the inserted and extracted images.

The model used for prediction is inspired from the model presented in [27]. We first perform data preparation, which includes image resizing to 224x224 and normalization, to meet the model's input requirements and data augmentation. A convolutional neural network (CNN) is then built for classification purposes, using

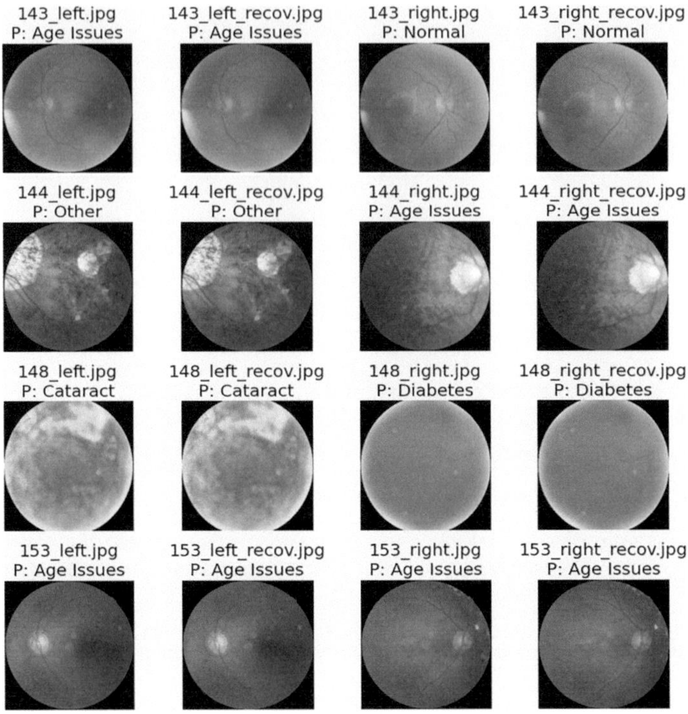

Fig. 8 Results of prediction after extraction

two convolutional blocks with batch normalization, max pooling and dropout for regularization. This is followed by a fully connected layer and an output layer with softmax activation for classification into 8 categories. The model is then combined with a pre-trained base model (EfficientNet) for fine-tuning. Finally, it is compiled using the Adam optimiser, with categorical cross-entropy as the loss function and accuracy as the tracking metric. The accuracy of the model is 74%. The precision of the model is 74% for cataract, 74% for normal, 76% for diabetic and 81% for other. Figure 8 shows the prediction results of the model before and after the images are inserted into the host. The predictions remained identical before and after embedding, confirming a 100% match across the entire test dataset. This indicates that the insertion and extraction processes do not alter the shape or intrinsic information of the data. Consequently, the analysis performed prior to the insertion of the dataset yields the same results as that performed after the extraction of the data from the host images.

4.6 Discussion

The process of securing and storing the medical data begins with RSA key pair generation, which establishes the cryptographic foundation for ensuring data authenticity. The RSA keys are used to sign both the medical data and the steganographed image, guaranteeing their integrity and authenticity throughout the data's lifecycle. Ganache, a local blockchain simulation tool, is employed to create a decentralized and immutable environment for securely storing the data. Blockchain's immutability ensures that once data is recorded, it cannot be altered or deleted, which adds a crucial layer of protection against tampering. In the data preparation phase, the medical images were resized and compressed using the SWT method. This step helps reduce the size of the secret image, making it more manageable for embedding while preserving its essential visual details. Once the image is compressed, it is embedded into the cover image using the LSB method. This process allows for efficient hiding of sensitive medical information while ensuring the visual quality of the cover image remains unaffected. To ensure confidentiality during the transmission and storage of the data, the system employs encryption using a session key generated by an authentication protocol. This approach has several advantages over using a static encryption key. A session key is unique to each communication session, meaning that even if the key is compromised in one session, the data from other sessions remains secure. This method also provides forward secrecy, ensuring that previous communications cannot be decrypted even if a future session key is compromised. In our system, both the steganographed image and the signed medical data are encrypted using this session key, ensuring that only authorized parties with the correct key can decrypt the data. This approach enhances data security during both transmission and storage.

After the encryption process, the encrypted data is stored on the blockchain through a custom smart contract named 'MedicalRecord'. This smart contract is responsible for storing important elements, including the hash of the encrypted medical data (dataHash), the hash of the steganographed image (imageHash), the RSA signatures (dataSignature and imageSignature), the unique sensor ID (sensorId), and the blockchain address of the entity that recorded the data (recordedBy). The contract also stores a timestamp indicating the precise time the data was stored. The use of blockchain ensures the immutability of the data, meaning that once recorded, the data cannot be altered or deleted. When the data is retrieved from the blockchain, the system verifies its integrity by comparing the stored hashes with newly computed ones from the retrieved data. If the hashes match, the system proceeds with decrypting the steganographed image and medical data using the session key. Once decrypted, the system verifies the RSA signatures of both the medical data and the steganographed image to confirm that they have not been tampered with since they were recorded. This step ensures the authenticity and integrity of the retrieved data. The next step in the process is extracting the secret image from the steganographed cover image. This is accomplished using the inverse LSB technique, which retrieves the bits of the secret image that were embedded in the cover image during the embedding phase. After retrieving the hidden bits, the system reassembles the approximation coefficients (cA) from the SWT, which represent the essential visual features of the image. The inverse SWT is then applied to reconstruct the original secret image, effectively reversing the wavelet decomposition and restoring the image with minimal loss of quality. Finally, both the secret image and the decrypted medical data are successfully recovered and ready for use. The high PSNR and NCC values demonstrate that the system is able to balance data security with maintaining the integrity and quality of the cover image. By employing encryption with a session key and storing the data immutably on the blockchain, the system guarantees that the sensitive medical information remains secure, confidential, and verifiable throughout its entire lifecycle. This robust combination of cryptographic techniques, steganography, and blockchain storage ensures that the system is highly secure, traceable, and reliable for managing sensitive medical data.

5 Conclusions

This work has introduced a secure and modular framework for managing sensitive medical data by integrating cryptography, steganography, blockchain technology, and AI-based validation. The use of RSA digital signatures, AES session-key encryption, and immutable smart contract storage ensures data confidentiality, integrity, and authenticity across the entire pipeline. The combination of Stationary Wavelet Transform (SWT) and Least Significant Bit (LSB) embedding allows for effective data concealment without degrading image quality, as confirmed by high PSNR and NCC scores. In addition, AI-based validation helps to ensure that the embedding process does not compromise the intrinsic information, enabling reliable automated medical analysis. The adoption of blockchain technology reinforces compliance with privacy regulations while enabling secure sharing of medical records, especially in the context of telemedicine and decentralized healthcare platforms.

While the system shows strong results, some limitations should be noted. The current evaluation has been conducted on a limited set of medical datasets, and testing on more diverse modalities (e.g., MRI, ECG) is needed to confirm generalizability. Furthermore, the cryptographic scheme may require adaptation to post-quantum standards in the future, to maintain security against emerging quantum threats. Future work will focus on enhancing the system's scalability and robustness. This includes the integration of real-time data from IoT medical devices, deployment on public blockchain platforms, and the use of AI-guided adaptive embedding techniques such as saliency-based steganography or GANs (Generative Adversarial Networks). These improvements will reinforce the system's relevance for practical, secure, and intelligent medical data management.

References

1. Abood, O., Guirguis, S.: A survey on cryptography algorithms. Int. J. Sci. Res. Publ. **8**, 495–516 (2018)
2. Weber, I., Lu, Q., Tran, A.B., Deshmukh, A., Gorski, M., Strazds, M.: A platform architecture for multi-tenant blockchain-based systems. In: 2019 IEEE International Conference on Software Architecture (ICSA), pp. 101–110 (2019)
3. Takaoğlu, M., Özyavaş, A., Ajlouni, N., Alshahrani, A., Alkasasbeh, B.: A novel and robust hybrid blockchain and steganography scheme. Appl. Sci. **11**(22), (2021)
4. Pustokhina, I.V., Pustokhin, D.A., Shankar, K.: Chapter 7-blockchain-based secure data sharing scheme using image steganography and encryption techniques for telemedicine applications. In: Jude, H.D., Gupta, D., Khanna, A., Khamparia, A. (eds.) Wearable Telemedicine Technology for the Healthcare Industry, pp. 97–108. Academic Press (2022)
5. Suresh, K.S., Kamalakannan, T.: Digital image steganography in the spatial domain using block-chain technology to provide double-layered protection to confidential data without transferring the stego-object. Int. J. Intell. Syst. Appl. Eng. **11**, 61–68 (2023)
6. Subramanian, N., Elharrouss, O., Al-ma'adeed, S., Bouridane, A.: Image steganography: a review of the recent advances. IEEE Access 1–1 (2021)
7. Abikoye, O., Ogundokun, R.: Efficiency of LSB steganography on medical information. Int. J. Electr. Comput. Eng. (IJECE) **11**, 4157 (2021)
8. Ngom, A., Djimnaibeye, S., Ngom, N.F., Sidibé, S., Niang, O.: A new wavelet based steganography method for securing medical data. In: Mambo, A.D., Gueye, A., Bassioni, G. (eds.) Innovations and Interdisciplinary Solutions for Underserved Areas, pp. 132–143. Springer Nature Switzerland, Cham (2022)
9. Alhatami, E., Bhatti, U., Huang, M., Feng, S.: Review and Enhancement of Discrete Cosine Transform (DCT) for Medical Image Fusion, pp. 89–97 (2023)
10. Prasad, P., Subramani, S., Bhavana, V., Krishnappa, H.: Medical image fusion techniques using discrete wavelet transform. In: 2019 3rd International Conference on Computing Methodologies and Communication (ICCMC), pp. 614–618 (2019)
11. Chaudhary, A., Sharma, A., Gupta, N.: Designing a secured framework for the steganography process using blockchain and machine learning technology. Int. J. Intell. Syst. Appl. Eng. **11**, 96–103 (2023)
12. Pramanik, S., Ghosh, R., Ghonge, M., Narayan, V., Sinha, M., Pandey, D., Samanta, D.: A Novel Approach Using Steganography and Cryptography in Business Intelligence, pp. 192–217 (2020)
13. Sothe, C., Schimalski, M.B., La Rosa, L.E.C., Castro, J.D.B., Feitosa, R.Q., Dalponte, M., Lima, C.L., Liesenberg, V., Miyoshi, G.T., De Almeida, C.M., Tommaselli, A.M.G.: Comparative performance of convolutional neural network, weighted and conventional support vector machine and random forest for classifying tree species using hyperspectral and photogrammetric data. GIScience Remote. Sens. **57**(3), 369–394 (2020)
14. Raja, V., Suresh, K.S.: Deep steg block: deep learning-enhanced steganography for secure communication in IoT devices using blockchain. Educ. Adm. Theory Pract. **30**, 2958–2972 (2024)
15. Chao, Z., Duan, X., Jia, S., Guo, X., Liu, H., Jia, F.: Medical image fusion via discrete stationary wavelet transform and an enhanced radial basis function neural network. Appl. Soft Comput. **118**, 108542 (2022)
16. Aufa, F.J., Endroyono, Affandi, A.: Security system analysis in combination method: RSA encryption and digital signature algorithm. In: 2018 4th International Conference on Science and Technology (ICST), pp. 1–5 (2018)
17. Shete, P., Kohle, S.: Image encryption using AES algorithm: study and evaluation. Int. J. Res. Appl. Sci. Eng. Technol. **10**, 1134–1137 (2022)
18. Djimnaibeye, S., Ngom, A., Sow, D.: Post-quantum cryptography for healthcare: a number theory based two-factor mutual authentication and key exchange protocol over lattices for TMIS. Adv. Appl. Discrete Math. **41**, 1–26 (2023)

19. Djimnaibeye, S., Ngom, A., Tchappi, I., Hassan, B. M., Najjar, A.: Improvement of cloud-assisted identity-based anonymous authentication and key agreement protocol for secure WBAN. In: Tchakounté, F., Atemkeng, M., Rajagopalan, R.P. (eds.) Safe, Secure, Ethical, Responsible Technologies and Emerging Applications-First EAI International Conference, SAFER-TEA 2023, Yaoundé, Cameroon, Proceedings. Lecture Notes of the Institute for Computer Sciences, Social Informatics and Telecommunications Engineering. vol. 566, pp. 58–70. Springer (2023)
20. Taherdoost, H.: Smart contracts in blockchain technology: a critical review. Information **14**, 117 (2023)
21. McKinney, W.: Python for Data Analysis: data Wrangling with pandas. NumPy, and Jupyter, p. 579. O'Reilly Media (2022)
22. Chollet, F.: Deep Learning with Python, 2nd edn. ISBN 978-1617296864 (2021)
23. U.S. Department of Health and Human Services: Health insurance portability and accountability act of 1996 (hipaa). https://www.hhs.gov/hipaa/for-professionals/index.html (1996). Accessed 06 Aug 2024
24. European Parliament and Council of the European Union: Regulation (EU) 2016/679 of the European parliament and of the council of 27 april 2016 on the protection of natural persons with regard to the processing of personal data and on the free movement of such data (general data protection regulation, gdpr). https://eur-lex.europa.eu/eli/reg/2016/679/oj (2016). Accessed 06 Aug 2024
25. Kaggke: Lung cancer mortality datasets v2. https://www.kaggle.com/datasets/masterdatasan/lung-cancer-mortality-datasets-v2
26. Kaggke: Ocular disease recognition (odir). https://www.kaggle.com/datasets/andrewmvd/ocular-disease-recognition-odir5k
27. Kaggke: Ocular disease recognition (odir) model. https://www.kaggle.com/code/kareemabdelhamed/ocular-diseases-deep-analysis-deep-learning-model

Smart Edge-AI Models for Real-Time Contamination Detection in Livestock Water Supplies

Babacar Mbaye Faye, Fatou Diop, and Ibrahima Niang

Abstract Water quality is crucial for health and productivity in livestock operations, as contaminated water can increase disease transmission risks and lead to livestock losses, directly impacting agricultural performance and veterinary costs. Traditional water quality monitoring methods, such as laboratory tests, are costly, slow, and often unsuitable for real-time detection, limiting prompt corrective actions, particularly in agricultural settings. Emerging technologies, including AI, IoT, and Edge Computing, now enable rapid and precise contaminant detection through multi-parametric sensors and machine learning algorithms, significantly reducing detection latency and enhancing the reliability of health monitoring systems. This paper proposes an intelligent model based on artificial intelligence and Edge-AI computing for real-time detection and prediction of contamination at livestock water points. By integrating multi-parameter biosensors, our layered architecture (Perception, Edge-AI, Network, Cloud) enables optimal resource management with reduced latency and increased accuracy compared to traditional methods. Experiments demonstrate fast detection of pathogenic contaminants, thus enhancing disease prevention in agricultural environments. This innovative model, which combines IoT, artificial intelligence, and autonomous floating platforms, also has potential for environmental applications.

Keywords Edge computing · Internet of things (IoT) · Biosensors · Artificial intelligence (AI) · Edge-AI · Machine learning · Smart agriculture · Real-time monitoring · Contaminant detection · Water quality · Predictive modeling

---

B. M. Faye (✉) · F. Diop · I. Niang
Dakar Computer Science Laboratory, Cheikh Anta Diop University, Dakar, Senegal
e-mail: babacarmbaye.faye@ucad.edu.sn

F. Diop
e-mail: fatou110.diop@ucad.edu.sn

I. Niang
e-mail: ibrahima1.niang@ucad.edu.sn

© The Author(s), under exclusive license to Springer Nature Switzerland AG 2026

C. T. Gueye et al. (eds.), *Mathematics, Computer Science, Cybersecurity and Artificial Intelligence*, Springer Proceedings in Mathematics & Statistics 536,
https://doi.org/10.1007/978-3-032-17666-0_4

1 Introduction

Water is critical to livestock health and agricultural productivity. Contaminated water increases the risk of disease transmission and animal loss, which directly affects farm performance and raises veterinary costs [1]. Preventing waterborne disease outbreaks is thus essential to safeguard both livestock and the communities that rely on agricultural systems.

Traditional monitoring methods, such as manual sampling and laboratory tests, are accurate but often **costly, slow, and impractical** for real-time detection, particularly in remote agricultural settings where lab infrastructure is limited. Pathogens like E. coli or Salmonella pose serious health threats, making **rapid and continuous detection** a pressing need [2].

Emerging technologies such as artificial intelligence (AI), the Internet of Things (IoT), and edge computing now offer viable alternatives. By coupling multi-parametric sensors with machine learning algorithms, these systems enable **on-site, real-time analysis** of water quality. This drastically reduces latency, enhances monitoring accuracy, and facilitates early intervention before clinical symptoms emerge [3].

In this work, we introduce a smart monitoring model for livestock water sources that integrates Edge-AI, biosensors, and distributed computing. This architecture aims to detect contamination proactively, improve animal health, and ensure efficient water management through intelligent resource optimization [4].

This paper is structured as follows. Section 2 presents related work by highlighting traditional approaches and their limitations. Section 3 describes our innovative water source monitoring architecture that integrates wireless sensor networks, IoT, and advanced machine learning techniques. Section 4 explains our validation methodology through simulations and case studies. Then, Sect. 5 discusses the results and compares our method's effectiveness to that of existing systems. Finally, Sect. 6 concludes by summarizing the main contributions of this study and proposes future research directions for monitoring water quality.

2 Related Work

Detection and prevention of waterborne diseases in livestock environments have gained significant momentum with the development of smart sensors and AI. While real-time monitoring systems have advanced, **challenges remain regarding accuracy, scalability, and real-time data processing**.

Traditional approaches based on manual sampling and lab analysis are reliable but **too slow for rapid interventions**, especially in remote or large-scale environments [5]. Biosensor based and intelligent sensor network systems have shown promise for real-time pathogen detection [6, 7], yet they still face **performance limitations**, particularly in high-scale deployments.

Recent studies have introduced AI to monitor water usage and detect anomalies, but most rely on **cloud-based infrastructures**, which are **unsuitable for rural areas** with poor connectivity [8]. Moreover, many AI systems in agriculture focus on resources management and animal health rather than real-time contamination detection [9].

Edge computing offers a promising alternative by enabling **on-site data analysis**, reducing latency and dependency on central servers. Although Edge-AI has been applied in fields like air and coastal water monitoring [10, 11], its use in **livestock water quality** remains scarce. Our approach fills this gap by integrating Edge-AI with multi-parametric biosensors for **local, real-time contamination detection**.

Unlike cloud-dependent models, our system provides **continuous on-site monitoring** and immediate response capabilities through embedded AI. By minimizing transmission delays and leveraging predictive analytics, our architecture anticipates potential outbreaks and underexplored dimensions in livestock water monitoring [12, 13].

3 Design Architecture

Our proposed architecture adopts a **multi-layered design** integrating IoT and Edge-AI technologies to enable fast, on-site contamination detection. It is structured into four key layers: **Perception (IoT), Edge-AI, Network, and Cloud**.

This distributed approach enhances responsiveness and energy efficiency, catering to the needs of remote agricultural environments where network infrastructures are limited.

Perception IoT layer This layer forms the interface with the physical environment. It consists of intelligent floating platforms equipped with biosensors and environmental sensors that capture real-time water quality parameters such as pH, turbidity, temperature, conductivity, and dissolved oxygen. Each platform has **local pre-processing capabilities** for outlier filtering and data compression before transmission. To conserve energy and enhance network stability, **autonomous clustering algorithms** dynamically elect a cluster head based on node energy and signal strength. Communication between sensors and platforms uses **low-power, long-range protocols** like ZigBee, LoRa, and BLE, and is secured using **elliptic curve cryptography**. Platforms ensure continuous data acquisition even under poor network conditions through adaptive caching mechanisms.

- **Sensors and data collection**: the sensors used include biosensors and environmental sensors to detect various contamination parameters. Types of sensors include **pH** (indicating the acidity or alkalinity of the water, crucial for identifying conditions conducive to pathogenic contaminant growth), **turbidity** (measuring water clarity, with high turbidity possibly indicating suspended organic particles), **temperature** (an essential indicator of water health, as higher temperatures promote

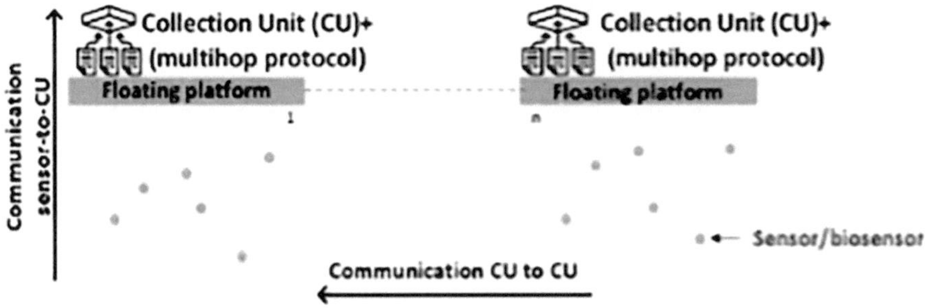

Fig. 1 Floating collector system architecture

the growth of certain harmful microorganisms), **conductivity** (assessing dissolved salt concentration as an indirect indicator of overall water quality), and biosensors. These sensors are networked within the floating platforms, ensuring homogeneous and adaptable coverage over large water bodies. The collected data is transmitted to collection units for initial pre-processing.
- **Intelligent floating platforms**: equipped with underwater biosensors and environmental sensors, these mobile platforms optimize data collection across large water bodies. They communicate with each other to form autonomous clusters, enabling flexible and decentralized management.
- **Data pre-processing in collection units**: the floating platforms remove outliers and compress data through initial processing to reduce latency. An adaptive caching mechanism ensures data security and availability in case of communication disruptions.
- **Communication protocols and encryption**: communication between sensors and platforms uses technologies like ZigBee, LoRa, and BLE (Bluetooth Low Energy), offering low energy consumption and extended range, protected by robust security protocols such as elliptic curve encryption (Fig. 1).

Edge-AI layer This layer handles **real-time processing and decision-making** close to the data source. It includes **Edge devices**, **AI gateways**, and **Edge servers** with neuromorphic processors capable of executing ML tasks such as:

- **Pattern recognition** via CNNs to detect contamination signatures.
- **Anomaly detection** using SVMs or threshold-based methods.
- **Pathogen classification** via supervised learning.

Edge gateways aggregate pre-processed data, adjust sampling rates dynamically, and trigger alerts when necessary. This **distributed intelligence** minimizes latency and optimizes energy usage by reducing cloud interactions.

Cloud layer The cloud supports **long-term storage, advanced analytics, and user interface access**. It performs:

Fig. 2 Edge-AI layer

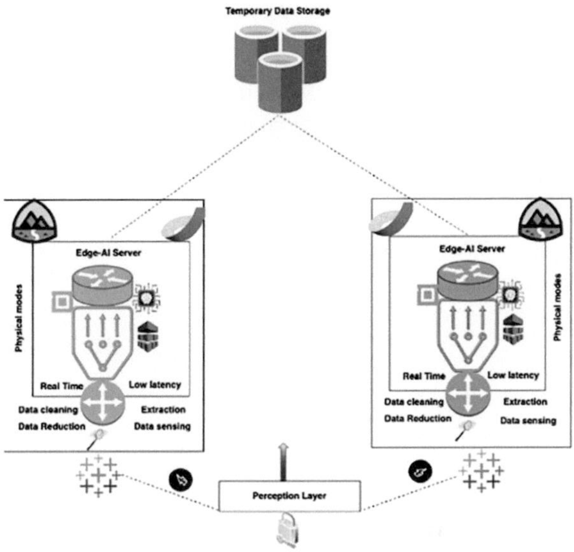

- **Historical data archiving** and trend analysis (via ARIMA or LSTM).
- **Risk forecasts** and contamination predictions.
- **Dashboard** reporting and system configuration tools for stakeholders (Fig. 2).

3.1 Network Layer

Implementing such architecture in rural areas poses challenges in connectivity, energy efficiency, and data reliability. Communication protocols between different layers play a central role in network fluidity and meeting environmental constraints.

- **ZigBee/LoRaWan** connect underwater sensors to surface units [14].
- **LTE/5G** links surface units to Edge gateways [15].
- **Long-range Wi-Fi or cellular** transmits processed data to the cloud. Each device is authenticated via **digital certificates**, ensuring **end-to-end encryption** throughout the stack.

This layered and modular design ensures scalability, responsiveness, and energy efficiency, making it well-suited for rural or isolated agricultural contexts where network infrastructure is limited.

4 Methodology

To validate our system, we will use a layered formalism (as a methodology) that defines each layer as a mathematical entity, specifying inputs, outputs, and operations. This formalism will use notation to represent components, data flow, and processing logic through each layer.

Let the system architecture S be defined by a layered model with three main layers: $\mathcal{L}_P$ (Perception Layer), $\mathcal{L}_E$ (Edge-AI Layer) and $\mathcal{L}_C$ (Application Layer in the Cloud).

$$S = (\mathcal{L}_P, \mathcal{L}_E, \mathcal{L}_C)$$

Each layer is responsible for specific operations, formally defined as follows:

A. Perception layer $\mathcal{L}_P$

Definition: the perception layer $\mathcal{L}_P$ is responsible for collecting data from the environment. It consists of n floating collection units $\mathcal{F}_i$, each equipped with sensors and biosensors for real-time data acquisition.

$$\mathcal{L}_P = \{\mathcal{F}_i \mid i = 1, 2, \ldots, n\}$$

Each floating collection unit $\mathcal{F}_i$ is defined as follows:

- **Sensors** S_{ij}: measure parameter j (e.g., pH, turbidity) and produce measurements $X_{ij}(t)$ at time t.
- **Data pre-processing** P_i: applies transformations such as outlier removal and compression to the raw data $X_{ij}(t)$.
- **Communication module** C_i: relays pre-processed data to the Edge-AI Layer $\mathcal{L}_E$.

For each floating collection unit $\mathcal{F}_i$, let:

$$X_i(t) = \{X_{ij}(t) \mid j = 1, 2, \ldots, m\}$$

where $X_i(t)$ represents measurements for all parameters at $\mathcal{F}_i$.

- $Y_i(t) = P_i(X_i(t))$: pre-processed data output from P_i.
- The communication module transmits $Y_i(t)$ to the Edge-AI Layer $\mathcal{L}_E$.

B. Edge-AI Layer $\mathcal{L}_E$

Definition: the Edge-AI layer $\mathcal{L}_E$ processes incoming data from $\mathcal{L}_P$, performs anomaly detection, and generates alerts. It comprises gateways G_k and Edge-AI servers E_k.

$$\mathcal{L}_E = \{G_k, E_k \mid k = 1, 2, \ldots, p\}$$

Each gateway G_k and Edge-AI server E_k has distinct roles:

The **Gateway** G_k aggregates and groups data from collection units $\mathcal{F}_i$. Let $Y_{G_k}(t) = \sum_{i=1}^{n} Y_i(t)$ denote the aggregated data at gateway G_k, which is then transmitted to the Edge-AI server E_k.

The **Edge-AI Server** E_k performs various tasks, including **Anomaly Detection** A_k, which detects anomalies in $Y_{G_k}(t)$ using thresholds or machine learning models. Additionally, **Alert Generation** R_k classifies contamination levels and generates alerts based on the results of anomaly detection.

For each Edge-AI server E_k, let $A_k(t) = A_k(Y_{G_k}(t))$ represent the anomaly status (binary or continuous) for each parameter. Subsequently, $R_k(t) = R_k(A_k(t))$ represents alerts generated with classifications, such as low, medium, or high risk. The Edge-AI server E_k transmits $R_k(t)$ to the application layer in the cloud $\mathcal{L}_C$ for long-term storage and analysis.

C. Application layer in the cloud $\mathcal{L}_C$

Definition: the application layer in the cloud $\mathcal{L}_C$ performs long-term data storage, trend analysis, and supports monitoring applications.

$$\mathcal{L}_C = \{\mathcal{D}, \mathcal{T}, \mathcal{M}\}$$

The components of $\mathcal{L}_C$ include:

The **Data Storage** $\mathcal{D}$ component stores historical data $Y_i(t)$, aggregated data $Y_{G_k}(t)$, and alerts $R_k(t)$ for subsequent analysis.

The **trend analysis** $\mathcal{T}$ component calculates trends $T_i(t)$ for each parameter i using time-series models, such as ARIMA or LSTM, and generates predictive insights $P_i(t)$ for early contamination warnings.

The **monitoring and applications** $\mathcal{M}$ component provides user interfaces to monitor data $Y_i(t)$ and alerts $R_k(t)$. It also allows users to view historical trends $T_i(t)$ and predictions $P_i(t)$ to support decision-making.

D. Network communication

Define network communication paths as a set of communication protocols C_N used for data exchange:

Within the perception layer, the **sensor to collection Unit** C_N^{SC} communication typically uses a low-power protocol such as Zigbee.

From the perception layer to the Edge-AI layer, the **collection unit to gateway** C_N^{CG} communication relies on LoRaWAN or cellular networks to support long-distance transmission.

From the Edge-AI layer to the application layer in the cloud, the **Edge-AI Server to Cloud** C_N^{EC} communication uses LTE/5G for reliable, high-speed data transfer.

E. Formal data flow

Data collection involves the process $X_i(t) \rightarrow Y_i(t) \quad \forall i \in \mathcal{L}_P$, where $X_i(t)$ represents the raw data collected by each sensor i in the perception layer $\mathcal{L}_P$, and $Y_i(t)$ is the pre-processed or filtered data ready for aggregation.

Aggregation and anomaly detection follows with $Y_i(t) \rightarrow Y_{G_k}(t) \rightarrow A_k(t) \rightarrow R_k(t) \quad \forall k \in \mathcal{L}_E$. Here, $Y_{G_k}(t)$ denotes the aggregated data at each gateway G_k in the Edge layer $\mathcal{L}_E$, $A_k(t)$ represents the anomaly status derived from analyzing $Y_{G_k}(t)$, and $R_k(t)$ is the risk classification or alert generated based on detected anomalies.

Long-term analysis and forecasting proceeds as $R_k(t) \rightarrow \mathcal{D} \rightarrow \mathcal{T} \rightarrow \mathcal{M}$. In this phase, $R_k(t)$ is stored in the data repository $\mathcal{D}$ for historical tracking, which is then used for trend analysis $\mathcal{T}$ and forecasting future contamination risks. The results $\mathcal{M}$ are available for monitoring and decision

5 Mathematical Model for Real-Time Contamination Detection

For real-time contamination detection in water sources, we employ a combination of statistical techniques, anomaly detection methods, and time-series forecasting. This model is based on several key components: parameter monitoring, anomaly detection, multi-parametric correlation, adaptive threshold adjustment, and contamination forecasting.

A. Parameter monitoring and threshold-based detection

For each contamination parameter X_i (e.g., pH, turbidity, dissolved oxygen), let:

- $X_i(t)$: Real-time value of parameter i at time t
- T_i: Threshold for parameter i, defined by historical data or regulatory standards
- μ_i: Reference mean for parameter i, calculated from historical data
- σ_i: Standard deviation for parameter i.

A parameter is flagged as abnormal if:

$$|X_i(t) - \mu_i| > k \cdot \sigma_i$$

where k is a sensitivity constant (typically 2 or 3) chosen according to the desired confidence level for anomaly detection.

B. Anomaly detection with statistical and machine learning models

Z-Score for anomaly detection For each parameter X_i, calculate a Z-score:

$$Z_i(t) = \frac{X_i(t) - \mu_i}{\sigma_i}$$

If $|Z_i(t)| > k$, then $X_i(t)$ is flagged as an anomaly for parameter i.

Isolation forest or autoencoder For more sophisticated anomaly detection, an *Isolation forest* or *Autoencoder* model can be trained on historical data to identify abnormal patterns:

- Let $f(X)$ be the anomaly score function of the *Isolation forest* or *Autoencoder*, where $f(X) \in [0, 1]$.
- A parameter is classified as abnormal if $f(X) > \alpha$, where α is a predefined anomaly threshold (e.g., 0.5).

C. Multi-parameter correlation analysis

Some contamination events involve anomalies in multiple parameters. A correlation matrix C can be defined to analyze relationships between parameters:

$$C = \begin{bmatrix} 1 & \rho_{12} & \rho_{13} & \cdots & \rho_{1n} \\ \rho_{21} & 1 & \rho_{23} & \cdots & \rho_{2n} \\ \vdots & \vdots & \ddots & \vdots & \vdots \\ \rho_{n1} & \rho_{n2} & \rho_{n3} & \cdots & 1 \end{bmatrix}$$

where ρ_{ij} is the correlation coefficient between parameters i and j. If an anomaly is detected in both parameters i and j, and ρ_{ij} exceeds a threshold β, the event is flagged as a multi-parameter contamination event.

D. Dynamic threshold adjustment

The threshold T_i for each parameter can be adjusted based on historical data and seasonal trends. A moving average or exponentially weighted moving average (EWMA) can be used:

$$T_i(t) = (1 - \lambda)T_i(t - 1) + \lambda X_i(t)$$

where λ is the smoothing factor. This adjustment helps to adapt thresholds to gradual environmental changes.

E. Contamination classification and alert system

The algorithm classifies contamination levels based on the number and type of anomalies:

- Define a binary variable $A_i(t)$ for each parameter:

$$A_i(t) = \begin{cases} 1 & \text{if } |X_i(t) - \mu_i| > k \cdot \sigma_i \\ 0 & \text{otherwise} \end{cases}$$

- Sum the binary indicators to obtain a combined anomaly score:

$$S(t) = \sum_{i=1}^{n} A_i(t)$$

- Define alert levels based on $S(t)$: Low risk is indicated by $S(t) = 1$, medium risk by $S(t) = 2$, and high risk by $S(t) \geq 3$.

F. Predictive contamination forecasting

For contamination forecasting, a time-series model such as ARIMA or LSTM can be applied. Given historical values $X_i(t-1), X_i(t-2), \ldots$, we predict the next value $X_i(t+1)$:

$$X_i(t+1) = f(X_i(t), X_i(t-1), \ldots)$$

where f is the model function based on ARIMA or LSTM. The forecast helps proactively identify contamination risks.

Thus, we propose the following summary: **Anomaly detection**: $|X_i(t) - \mu_i| > k \cdot \sigma_i$ or $f(X) > \alpha$
Multi-parameter correlation: If $\rho_{ij} > \beta$ and both X_i and X_j are abnormal, the event is flagged as a correlated anomaly.
Adaptive threshold: $T_i(t) = (1 - \lambda)T_i(t-1) + \lambda X_i(t)$.
Alert levels: calculate $S(t) = \sum_{i=1}^{n} A_i(t)$ and classify the risk according to $S(t)$.
Forecasting: use ARIMA or LSTM to predict $X_i(t+1)$ for preventive contamination alerts.

This model provides a comprehensive mathematical framework to support real-time contamination detection, correlation, adaptive management, and predictive alerts at livestock water sources.

6 Real-Time Contamination Detection Algorithm

6.1 Parameter Monitoring and Threshold Definition

Contamination indicators: the algorithm monitors key parameters associated with water contamination, including pH levels, where deviations from neutral pH may indicate contaminants; turbidity, where high levels may suggest suspended particles or pathogens; dissolved oxygen (DO), with low DO levels potentially indicating organic pollution as contaminants consume oxygen; and temperature and conductivity, where variations may signal chemical or microbial contamination.

Threshold configuration: predefined thresholds are set for each parameter based on acceptable levels of water safety for livestock. These thresholds may vary according to local water quality regulations, livestock health standards, and environmental factors.

6.2 Data Preprocessing and Anomaly Detection

Noise reduction and data smoothing: sensor data often contains noise due to environmental variations. The algorithm applies smoothing techniques, such as moving averages or exponential smoothing, to reduce transient fluctuations that could trigger false alarms.

Anomaly detection model: the system first establishes a**baseline model** to define "normal" parameter ranges, derived from historical data or an initial training period. For **real-time anomaly detection**, the algorithm continuously compares incoming data to this reference, using statistical methods (Z-score, IQR) or machine learning models (isolation forest or autoencoders) to identify deviations from normal patterns.

6.3 Multi-parameter Correlation Analysis

Inter-parameter analysis: certain contamination events affect multiple parameters simultaneously. For example, an increase in turbidity and a decrease in DO levels may both indicate organic pollution. The algorithm correlates parameter changes to improve detection accuracy.
Pattern recognition: by recognizing multi-parameter patterns associated with specific contaminants, the algorithm can more accurately classify contamination types (chemical vs. biological).

6.4 Dynamic Threshold Adjustment

Adaptive thresholds: the algorithm dynamically adjusts parameter thresholds in response to seasonal variations, weather conditions, and historical trends to reduce false positives and enhance sensitivity.
Feedback mechanism: through a feedback loop, the algorithm can increase sampling frequency when abnormal readings are detected or reduce it during stable conditions to conserve energy.

6.5 Alert Generation and Response

Contamination classification: based on detected anomalies, the algorithm classifies contamination levels (e.g., low, medium, high risk) to guide response urgency.
Hierarchical alert system: **local alerts** are generated and transmitted locally to personnel or nearby control systems when contamination reaches critical levels,

providing immediate warnings. In addition, **remote alerts** are transmitted to a cloud application for centralized monitoring, if network conditions allow, enabling remote personnel to take action or analyze trends over time.
Automated actions: in case of severe contamination, the algorithm may trigger automated responses, such as shutting off the water source or activating a filtration system if integrated.

6.6 Predictive Maintenance and Contamination Forecasting

Machine learning-based forecasting models: the algorithm includes predictive models (time-series forecasting using ARIMA or LSTM) to anticipate contamination trends from historical data, which can aid in preventive maintenance and early interventions.
Predictive insights for resource planning: forecasting insights can inform livestock managers of potential contamination risks, allowing them to take proactive measures to secure alternative water sources or schedule water treatment.

6.7 Performance Indicators and Continuous Improvement

Accuracy and latency monitoring: the algorithm is continuously evaluated for its accuracy (true positive and false positive rates) and latency to ensure it meets real-time response requirements.
Algorithm refinement: based on performance data, the algorithm undergoes periodic updates or retraining to incorporate new environmental factors, contamination sources, or regulatory standards.

6.8 Pseudocode Outline for the Real-Time Contamination Detection Algorithm

This pseudocode captures the main logic steps, covering sensor data processing, anomaly detection, correlation analysis, threshold adjustment, alert generation, and response actions. The algorithm dynamically adapts to environmental changes,

Table 1 Notation legend for algorithm

Symbol	Description
$\mathcal{S}$	Real-time input stream of sensor data
θ	Adaptive threshold vector for contamination detection
$\mathcal{M}$	Baseline model trained on clean historical data
$\mathcal{P}$	Set of monitored water quality parameters
D	Preprocessed data vector from current cycle
x	Current reading of a parameter $p \in \mathcal{P}$
Γ	Set of anomalies detected and correlated in the current window
$\mathcal{A}$	Alert level: `Normal`, `medium risk`, or `high risk`
$\mathcal{R}$	Response actions triggered based on alert level

enhancing the accuracy and responsiveness of the contamination detection system (Table 1).

Algorithm 1: Real-time Contamination Detection Algorithm

```
Input: Sensor data stream S, adaptive thresholds θ, baseline model M
Output: Alert level A, response actions R
P ← {pH, Turbidity, DO, Temperature, Conductivity} ;      // Contamination parameters
Load M with historical normal values ;
θ ← ComputeThresholds(P) ;                                // Initialize adaptive thresholds
while S is active do
    D ← Preprocess(Fetch(S)) ;
    A ← Normal, R ← ∅, isAnomaly ← False ;
    foreach p ∈ P do
        x ← D[p] ;
        if x > θ[p] or DetectAnomaly(x, M) then
            Mark p as anomalous ;
            isAnomaly ← True ;
    if isAnomaly then
        Γ ← CorrelateParameters(D) ;
        A ← ClassifyContamination(Γ) ;
    UpdateThresholds(θ, D) ;
    if A = High Risk then
        TriggerAlert(both, Γ) ;
        R ← InitiateAutomatedResponse() ;
    else if A = Medium Risk then
        TriggerAlert(local, Γ) ;
        R ← RecommendInspection() ;
    else
        Log event for monitoring ;
    if PredictContaminationTrend(S) = High Risk then
        SchedulePreventiveMaintenance() ;
```

7 Results and Analysis

In this study, we generated baseline data for key water quality parameters pH, turbidity, dissolved oxygen, and temperature using a normal distribution centered on reference values typically observed in livestock environments. A contamination event was emulated by introducing an anomaly into the turbidity parameter, significantly increasing its value over a short duration. This setup allowed for testing the system's response under controlled yet realistic contamination conditions.

It is important to note that all evaluations in this work were conducted using simulated sensor data. These simulations were designed to reflect plausible contamination scenarios and environmental fluctuations in real-world agricultural contexts. While this simulation-based approach provides valuable insights into system behavior and parameter sensitivity, future work will involve real-world deployment and validation in operational livestock settings.

To detect anomalies in each parameter, we applied z-score normalization. Any data point with a z-score greater than 2 was flagged as an outlier. In addition, we implemented an adaptive thresholding technique based on an exponentially weighted moving average (EWMA), using a smoothing factor $\lambda = 0.1$. This method improves sensitivity to sudden shifts in the signal, such as contamination events with gradually shifting baselines.

The effectiveness of adaptive thresholds in monitoring parameter values and identifying critical anomalies corresponding to contamination events is demonstrated in Fig. 3.

To explore dependencies between variables, we computed a correlation matrix for pH, turbidity, dissolved oxygen, and temperature. This analysis reveals potential interrelationships that could enhance the predictive capabilities of future models.

A global contamination score was derived by summing the anomalies detected across all parameters. Based on this score, we defined four alert levels `Normal`, `Low Risk`, `Medium Risk`, and `High Risk` to classify the severity of contamination in real-time.

To forecast future contamination trends, an ARIMA model of order (1, 1, 1) was fitted to the turbidity signal, which is a strong indicator of water quality degradation. The prediction performance was evaluated using Mean Absolute Error (MAE) and Root Mean Square Error (RMSE) (Fig. 4).

These metrics validate the model's potential for reliable contamination forecasting, which is essential for early intervention.

8 Discussion

The results confirm the effectiveness of our Edge-AI framework for real-time detection and prediction of contaminants in livestock water sources. The proposed multi-layered architecture comprising Perception, Edge-AI, and Cloud components enables

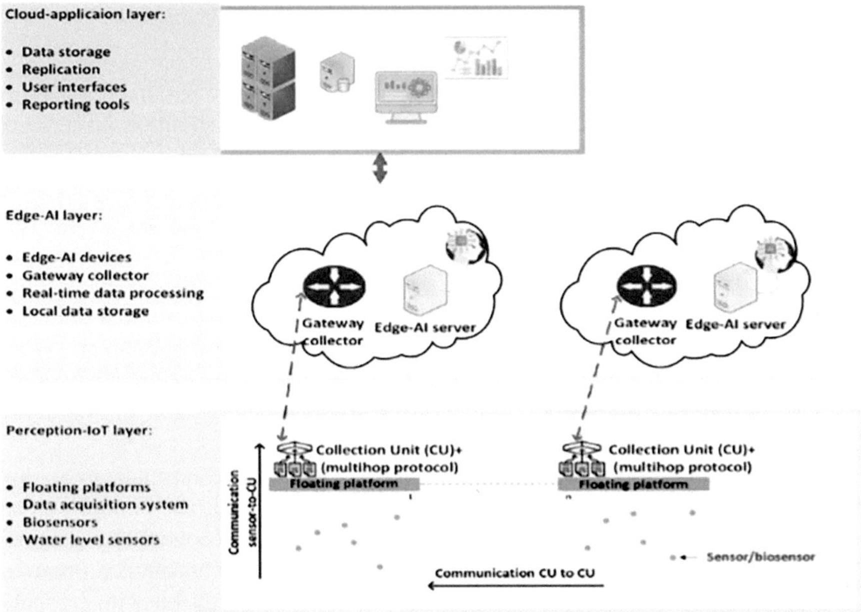

Fig. 3 Network architecture

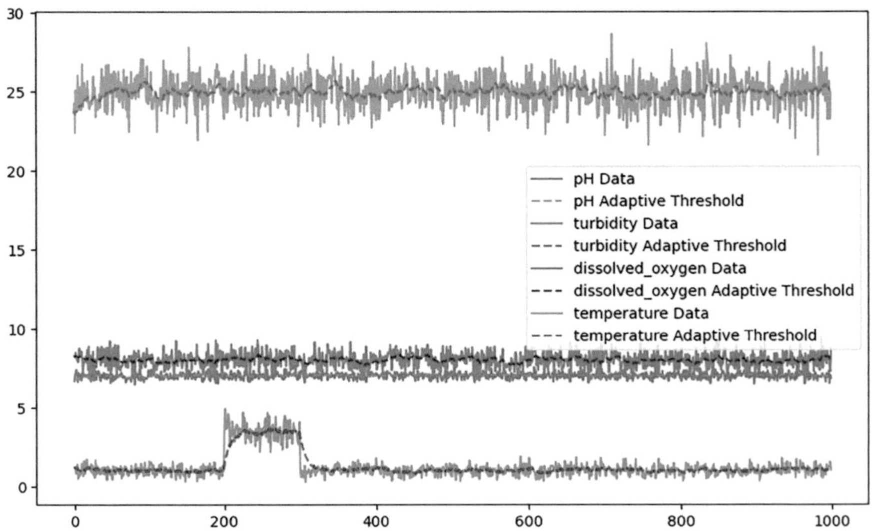

Fig. 4 Predictive modeling with ARIMA

efficient resource management while significantly reducing latency and improving detection accuracy compared to traditional centralized approaches.

Unlike conventional cloud-based solutions, our architecture minimizes dependence on external infrastructures, enhancing responsiveness and accessibility in rural areas where network connectivity is often limited. Localized Edge-AI analysis enables near-instantaneous responses to detected contamination events, thereby outperforming laboratory analysis and other conventional sensor systems in terms of reactivity.

However, the deployment of such a system in remote environments introduces several challenges, particularly regarding energy constraints and network coverage. Ensuring energy autonomy of the sensing platforms remains a critical priority. Strategies such as the integration of low-power sensors, adaptive sampling techniques, and renewable energy sources (e.g., solar panels) are essential to extend system longevity and ensure continuous operation under harsh environmental conditions.

Several avenues for future improvement can be explored to reinforce the robustness and trustworthiness of the system. For example, integrating blockchain mechanisms could enhance data security, traceability, and transparency, thereby increasing trust among stakeholders. Furthermore, this architecture shows potential for broader applications in environmental monitoring, including wetland preservation and agricultural water quality assessment, thereby contributing to sustainable ecosystem management.

By providing proactive and context-aware contamination alerts, the system supports early disease prevention and helps reduce livestock mortality, which is of paramount importance for farmers and water resource managers. This innovative approach contributes to enhanced animal health safety while reducing the economic burden associated with waterborne diseases and veterinary treatments.

9 Conclusion and Future Perspectives

This study presented an intelligent water quality monitoring framework for livestock water points, leveraging Edge-AI and multi-parametric biosensors for real-time detection and prediction of contamination events. The proposed layered architecture composed of Perception, Edge, and Cloud layers enables optimized resource usage, reduces latency, and improves responsiveness compared to traditional centralized approaches. By shifting critical processing tasks to the edge, the system reduces dependence on cloud infrastructure, thus enhancing autonomy and operational resilience in isolated agricultural environments.

The proposed solution was validated using simulated sensor data, which allowed us to emulate realistic contamination scenarios and evaluate the system's detection and prediction capabilities under controlled conditions. This simulation-based approach was essential to assess the effectiveness of our anomaly detection, adaptive thresholding, and forecasting mechanisms. Future work will focus on deploying the framework in real-world livestock environments to validate its robustness, responsiveness, and long-term reliability in operational contexts.

Several improvements can be considered to strengthen the system's capabilities. The integration of blockchain technologies could enhance data security, traceability, and transparency, fostering greater trust among users and stakeholders. Additionally, addressing energy constraints remains a key priority. Optimizing energy consumption in floating platforms and sensor units through the use of low-power devices, adaptive sampling, and renewable energy sources is essential to ensure long-term deployment in energy-scarce rural settings.

Finally, this architecture is flexible and extensible, and could be applied to a range of environmental monitoring domains, such as wetland preservation or sustainable water management, opening new perspectives for ecosystem protection and smart agriculture.

References

1. Zhu, M., et al.: A review of the application of machine learning in water quality evaluation. Eco-Environ. Health **1**(1), 107–116 (2022)
2. Sendra, S., et al.: LoRa-based network for water quality monitoring in coastal areas. Mob. Netw. Appl. **28**(1), 65–81 (2023)
3. Parra, L., et al.: Low-cost microcontroller-based multiparametric probe for coastal area monitoring. Sensors **23**(3), 1871 (2023)
4. Contreras-Castillo, J., et al.: SAgric-IoT: an IoT-based platform and deep learning for greenhouse monitoring. Appl. Sci. **13**(4), 1961 (2023)
5. Ramezanpour, K., et al.: Security and privacy vulnerabilities of 5G/6G and WiFi 6: survey and research directions from a coexistence perspective. Comput. Netw. **221**, 109515 (2023)
6. Saha, C., et al.: Pesticide sensing using electrochemical techniques: a comprehensive review. J. Iran. Chem. Soc. **20**, 243–256 (2022)
7. Umapathi, R., et al.: Advances in optical-sensing strategies for the on-site detection of pesticides in agricultural foods. Trends Food Sci. Technol. **119**, 69–89 (2022)
8. Massarelli, C., et al.: Dynamics of pesticides in surface water bodies by applying data mining to spatiotemporal big data. Ecol. Inform. **78**, 102342 (2023)
9. Singh, S., et al.: Real-time water quality monitoring of River Ganga (India) using internet of things. Ecol. Inform. **71**, 101770 (2022)
10. Liu, Y., et al.: Recent progress regarding electrochemical sensors for the detection of typical pollutants in water environments. Anal. Sci. **38**(1), 55–70 (2022)
11. Zhu, M., Wang, J., Yang, X., Zhang, Y., Zhang, L., Ren, H., Wu, B., Ye, L.: A review of the application of machine learning in water quality evaluation. Eco-Environ. Health **1**, 107–116 (2022)
12. Laha, S.R., Pattanayak, B.K., Pattnaik, S.: Advancement of environmental monitoring system using IoT and sensor: a comprehensive analysis. AIMS Environ. Sci. **9**, 771–800 (2022)
13. Fascista, A.: Toward integrated large-scale environmental monitoring using WSN/UAV/crowdsensing: a review of applications, signal processing, and future perspectives. Sensors **2022**, 22 (1824)
14. Zulkifli, C.Z., Garfan, S., Talal, M., Alamoodi, A.H., Alamleh, A., Ahmaro, I.Y.Y., Sulaiman, S., Ibrahim, A.B., Zaidan, B.B., Ismail, A.R., et al.: IoT-based water monitoring systems: a systematic review. Water **14**, 3621 (2022)

15. Umapathi, R., Park, B., Sonwal, S., Rani, G.M., Cho, Y., Huh, Y.S.: Advances in optical-sensing strategies for the on-site detection of pesticides in agricultural foods. Trends Food Sci. Technol. **119**, 69–89 (2022)
16. Contreras-Castillo, J., Guerrero-Ibañez, J.A., Santana-Mancilla, P.C., Anido-Rifón, L.: SAgric-IoT: an IoT-based platform and deep learning for greenhouse monitoring. Appl. Sci. **2023**, 13 (1961)
17. Singh, R.S.; Gelmecha, D.J.; Ayane, T.H.; Sinha, D.K. Chapter 4—Functional framework for IoT-based agricultural system. In: Abraham, A., Dash, S., Rodrigues, J.J.P.C., Acharya, B., Pani, S.K. (eds.) AI, Edge and IoT-Based Smart Agriculture, (Intelligent Data-Centric Systems). Academic Press, Cambridge, MA, USA (2022)
18. Sendra, S., Parra, L., Jimenez, J.M., Garcia, L., Lloret, J.: LoRa-based network for water quality monitoring in coastal areas. Mob. Netw. Appl. **28**, 65–81 (2023)
19. Pasolini, G.: On the LoRa chirp spread spectrum modulation: signal properties and their impact on transmitter and receiver architectures. IEEE Trans. Wirel. Commun. **21**, 357–369 (2022)
20. Nvs, B.; Saranya, P.L.: Chapter 18—Water pollutants monitoring based on internet of things. In: Devi, P., Singh, P., Kansal, S.K. (eds.) Inorganic Pollutants in Water. Elsevier, Amsterdam, The Netherlands (2020)
21. Sang, D., Cimetiere, N., Giraudet, S., Tan, R., Wolbert, D., Le Cloirec, P.: Online SPE-UPLC-MS/MS for herbicides and pharmaceuticals compounds' determination in water environment: a case study in France and Cambodia. Environ. Adv. **8**, 100212 (2022)
22. Schmidt, J.Q., Kerkez, B.: Machine learning-assisted, process-based quality control for detecting compromised environmental sensors. Environ. Sci. Technol. **57**, 18058–18066 (2023)
23. Li, R., Wang, Z., Zhang, Z., Sun, X., Hu, Y., Wang, H., Chen, K., Liu, Q., Chen, M., Chen, X.: Deep learning-based multicapturer SERS platform on plasmonic nanocube metasurfaces for multiplex detection of organophosphorus pesticides in environmental water. Anal. Chem. **94**, 16006–16014 (2022)
24. Lewis, S.W.: Spectroscopic techniques. In: Houck, M.M. (ed.) Encyclopedia of Forensic Sciences, 3rd edn. Elsevier, Oxford, UK (2023)

Generalized Subspace Subcodes in the Rank-Metric

Ousmane Ndiaye, Peter Arnaud Kidoudou, and Hervé Tale Kalachi

Abstract Rank-metric codes were studied by E. Gabidulin in 1985 after a brief introduction by Delsarte in 1978 as analogues of Reed-Solomon codes in the rank-metric, but based on linearized polynomials. They have found applications in many areas, including linear network coding and space-time coding. They are also used in cryptography to reduce the size of the keys compared to Hamming metric codes at the same level of security. However, some families of rank-metric codes suffer from structural attacks due to the strong algebraic structure from which they are defined. It therefore becomes interesting to find new code families in order to address these questions in the landscape of rank-metric codes. In this paper, we provide a generalization of Subspace Subcodes in Rank-metric introduced by Gabidulin and Loidreau. We also characterize this family by giving an algorithm which allows to have its generator and parity-check matrices based on the associated extended codes. We have also studied the specific case of Gabidulin codes whose underlying decoding algorithms are known. Bounds for the cardinalities of these codes, both in the general case and in the case of Gabidulin codes, are also provided.

Keywords Coding theory · Rank-metric · Gabidulin code · Cryptography · Shortened code · Punctured code · Subfield subcodes

O. Ndiaye (✉)
Université Cheikh Anta Diop de Dakar, FST, DMI, LACGAA, Dakar, Senegal
e-mail: ousmane3.ndiaye@ucad.edu.sn

P. A. Kidoudou
Faculty of Science and Technology, Université Marien Ngouabi (UMNG), Brazzaville, Republic of the Congo
e-mail: peter.kidoudou@umng.cg

H. Tale Kalachi
Computer Engineering Department, National Advanced School of Engineering of Yaoundé, University of Yaoundé 1, Yaoundé, Cameroon

© The Author(s), under exclusive license to Springer Nature Switzerland AG 2026
C. T. Gueye et al. (eds.), *Mathematics, Computer Science, Cybersecurity and Artificial Intelligence*, Springer Proceedings in Mathematics & Statistics 536,
https://doi.org/10.1007/978-3-032-17666-0_5

1 Introduction

Rank-metric codes were studied by Gabidulin in 1985 [9] after a brief introduction by Delsarte in 1978 [7], who gave a description based on finite fields and their properties. By construction, Gabidulin codes are analogues of Reed-Solomon codes in the rank-metric [30], since codewords are obtained by evaluation of $q-$polynomials on a support included in an extension of degree m of $\mathbb{F}_q$. Rank-metric codes have found applications in network coding [8], for example, when the transmitter and receiver are oblivious to the inner workings and topological network. Rank-metric codes have also used in the theory of space-time codes, introduced by Lu and Kumar from Gabidulin codes [22].

Current cryptographic systems mostly rely on number-theoretic problems such as integer factorization and discrete logarithms. However, these problems are all vulnerable to quantum algorithms [31]. The theory of error correcting codes is a serious candidate that offers perspectives to face quantum computers via the Syndrome Decoding problem in the Hamming metric or in the Rank-metric. They have been widely used in cryptography over the last few decades to provide cryptographic primitives for encryption, signature, hashing, or pseudo-random number generation. The main reason for using rank-metric codes in cryptography is the possibility of reducing the size of the keys compared to Hamming metric codes at the same level of security. Rank-metric-based cryptography began with the GPT cryptosystem and its variants [4, 12, 13, 21, 28, 29] based on Gabidulin codes [9], which are rank-metric analogues of Reed-Solomon codes. Unfortunately, as in the case of Reed-Solomon codes, the strong algebraic structure of these codes has been successfully exploited to attack the original GPT cryptosystem and its variants in a series of works initiated by Gibson and Overbeck [14, 15, 17, 20, 26, 27]. Recently, some public key cryptosystems based on rank-metric codes [23, 24] were candidates at the NIST competition for Post-Quantum Cryptography and made it to the second round. But again, they were all eliminated after the second round of the competition due to security defects. Considering the fact that the very old McEliece cryptosystem that uses Subspace Subcodes of Generalized Reed-Solomon codes is still secure and currently at the fourth round of the NIST competition, it becomes theoretically interesting to study Subspace Subcodes in the rank-metric.

The notion of Subspace Subcode is used to denote a Subcode whose components of each codeword belong to the same vector subspace. This notion was first introduced in the Hamming metric for Reed-Solomon codes by Hattori et al. [16]. A few years later, the same notion was introduced for rank-metric codes by Gabidulin and Loidreau with applications to cryptography [4, 11]. Recently, Berger et al. [3] proposed a generalization of these Subcodes in the Hamming metric by allowing the components of each codeword of the Subcode to be in different subspaces. This previous work was followed by an article of Berger et al. [2], proposing a cryptosystem based on quasi-cyclic Subcodes of Subspace Subcode of Reed-Solomon codes.

This paper introduces and characterizes the family of Generalized Subspace Subcodes of a rank-metric code by giving an algorithm which allows to construct their generator and parity-check matrices based on the associated extended codes. Bounds of the cardinalities of these codes both in the general case and in the case of Gabidulin codes are also provided as a generalization of the results obtained in [11]. The work ends with a focus on the specific case of Gabidulin codes whose decoding algorithms are known.

The rest of the document is organized as follows: we start in Sect. 2 with some preliminaries to identify the generalities related to rank-metric codes. Section 3 presents some results on Generalized Subspace Subcodes in the rank-metric and a new result on the lower bound of their size. We give our main results on Generalized Subspace Subcodes of Gabidulin codes in Sect. 4 by giving bounds of their size and an algorithm to construct directly a generator matrix. Section 5 deals with the parent Codes of Gabidulin RGSS[1] codes for decoding subcodes as an alternative to using the decoding algorithm of the supercode. In Sect. 6 we give some directions for their potential applications in cryptology and how it can reduce the key sizes. Finally, Sect. 7 concludes the paper and gives some perspectives.

2 Preliminaries

2.1 Rank-Metric Codes

The *rank weight* $w_R(\mathbf{x})$ of a word $\mathbf{x} = (x_1, \ldots, x_n) \in \mathbb{F}_{q^m}^n$ in an extension field $\mathbb{F}_{q^m}/\mathbb{F}_q$ is defined by the maximal number of its elements that are linearly independent over the base field $\mathbb{F}_q$, where m and n are positive integers and q a power of a prime number. The *rank distance* d_R between two words is defined by the rank weight of their difference, i.e., $d_R(\mathbf{x}, \mathbf{y}) = w_R(\mathbf{x} - \mathbf{y})$. It is well known that d_R has the properties of a metric on $\mathbb{F}_{q^m}^n$.

Definition 1 Let $\mathbf{x} = (x_1, \ldots, x_n) \in \mathbb{F}_{q^m}^n$ and $\{b_1, b_2, \ldots, b_m\}$ a basis of $\mathbb{F}_{q^m}$ over $\mathbb{F}_q$. We can write $x_j = \sum_{i=1}^m x_{ij} b_i \in \mathbb{F}_{q^m}$ for each $j = 1, \cdots, n$ with $x_{ij} \in \mathbb{F}_q$. The rank weight $w_R(\mathbf{x})$ of $\mathbf{x}$ is defined as the rank of the matrix $M_{\mathbf{x}} = (x_{ij}) \in \mathbb{F}_q^{m \times n}$.

In the sequel, a linear code of length n and dimension k (also called $[n, k]-$linear code or simply $[n, k]-$code) over $\mathbb{F}_{q^m}$ will denote a $k-$dimensional subspace of the $n-$dimensional vector space $\mathbb{F}_{q^m}^n$. A rank-metric code is then a linear code endowed with the above metric, called rank-metric. Given a rank-metric code $\mathcal{C} \subset \mathbb{F}_{q^m}^n$, its minimum rank distance is

$$d_R(\mathcal{C}) = \min_{\mathbf{c}_1 \neq \mathbf{c}_2 \in \mathcal{C}} w_R(\mathbf{c}_1 - \mathbf{c}_2) = \min_{\mathbf{c} \in \mathcal{C}, \mathbf{c} \neq \mathbf{0}} w_R(\mathbf{c}).$$

If $\mathcal{C} \subset \mathbb{F}_{q^m}^n$ is a $k-$dimensional rank-metric code with minimum rank distance $d_R(\mathcal{C})$, it is said to be a $[n, k, d = d_R(\mathcal{C})]_{q^m}-$code. The parameters of a $[n, k, d =$

[1] Rank Generalized Subspace Subcodes.

$d_R(\mathcal{C})]_{q^m}-$code are related by an equivalent of the Singleton bound for the rank distance, see [9]:

$$Card(\mathcal{C}) \le q^{\min(m(n-d+1),n(m-d+1))}$$

where $Card$ is the cardinality of $\mathcal{C}$. Furthermore, a code satisfying the equality $Card(\mathcal{C}) = q^{\min(m(n-d+1),n(m-d+1))}$ is called a Maximum Rank Distance (MRD) code.

2.2 Punctured and Shortened Codes

Punctured and shortened codes of a given code are very important derivative codes to characterize certain subcodes, but also to mount structural attacks against code-based cryptosystems [6]. We are going to use them on extended codes to characterize Subspace Subcodes in the rank-metric. We recall here their definitions and some needed properties.

Definition 2 (Punctured code)
Let $\mathcal{C} \subset \mathbb{F}_q^n$ be a linear code, $\mathcal{I} \subseteq \{1, .., n\}$ a set of coordinate positions.

- Given a vector $\mathbf{y} \in \mathbb{F}_q^n$, the punctured vector $\mathbf{y}\mathcal{I}$ of $\mathbf{y}$ on $\mathcal{I}$ is defined by $\mathbf{y}\mathcal{I} := (y_i)_{i\in\{1,\dots,n\}\setminus\mathcal{I}}$.
- The punctured code $\mathcal{C}\mathcal{I}$ of $\mathcal{C}$ on $\mathcal{I}$ is the code of length $n - |\mathcal{I}|$ defined by $\mathcal{C}\mathcal{I} = \{\mathbf{c}\mathcal{I} \mid \mathbf{c} \in \mathcal{C}\}$.

In this paper, we will sometimes use the puncturing operation on matrices as follows. For a matrix $\boldsymbol{M} \in \mathbb{F}_q^{m\times n}$, the punctured matrix of $\boldsymbol{M}$ on $\mathcal{I}$ is the matrix $\boldsymbol{M}\mathcal{I} \in \mathbb{F}_q^{m\times(n-|\mathcal{I}|)}$ obtained by puncturing all the row vectors of $\boldsymbol{M}$ on $\mathcal{I}$.

Proposition 1 *Let $\mathcal{C} \subset \mathbb{F}_q^n$ be a $[n, k, d]-$code, $\mathcal{I} \subseteq \{1, \dots, n\}$. Then $\mathcal{C}\mathcal{I}$ is an $[n - |\mathcal{I}|, k', d']-$code such that:*

$$k - |I| \le k' \le k \text{ and } d - |I| \le d' \le d.$$

Definition 3 (Shortened Code)
Let $\mathcal{C} \subset \mathbb{F}_q^n$ be a linear code, $\mathcal{I} \subseteq \{1, \dots, n\}$ a set of coordinates and $\mathcal{C}' = \{c \in \mathcal{C} \mid c_i = 0,\ \forall\, i \in \mathcal{I}\}$. The Shortened code of $\mathcal{C}$ on $\mathcal{I}$ denoted by $\mathcal{C}\mathcal{I}$ is the punctured code of $\mathcal{C}'$ on $\mathcal{I}$ i.e., $\mathcal{C}\mathcal{I} = \mathcal{C}'\mathcal{I}$.

Proposition 2 *Let $\mathcal{C} \subset \mathbb{F}_q^n$ be a $[n, k, d]-$code and $\mathcal{I} \subseteq \{1, \dots, n\}$. Then $\mathcal{C}\mathcal{I}$ is a $[n - |\mathcal{I}|, k', d']-$code satisfying:*

$$k - |I| \le k' \le k \text{ and } d \le d'.$$

Theorem 1 (Link between Shortened and Punctured Codes [[18], Theorem 1.5.7])
Let $\mathcal{C} \subset \mathbb{F}_q^n$ be a $[n, k, d]-$code and $\mathcal{I} \subseteq \{1, \dots, n\}$. We have

1. $(\mathcal{C}^{\perp})\mathcal{I} = (\mathcal{C}\mathcal{I})^{\perp}$
2. $(\mathcal{C}\mathcal{I})^{\perp} = (\mathcal{C}^{\perp})\mathcal{I}$.

This link will allow us to have a generator matrix of a shortened code (subcode) of a code from the punctured code of its dual code.

3 Generalized Subspace Subcodes in the Rank-Metric

Subspace Subcodes in the rank-metric were introduced by Gabidulin and Loidreau in [11], where decoding methods were also proposed. In this section, we give a generalization of Subspace Subcodes in the rank-metric.

Definition 4 ([11]) Let $\mathcal{C}$ be a $[n, k, d = d_R(\mathcal{C})]_{q^m}$−code, and V a $\mathbb{F}_q$−vector subspace of $\mathbb{F}_{q^m}$ with dimension $s \leq m$. The Subspace Subcode of $\mathcal{C}$, with respect to V, is the $\mathbb{F}_q$−vector space $\mathcal{C} \cap V^n$.

The elements of $\mathcal{C} \cap V^n$ are codewords whose components lie in the alphabet formed by the subspace V. In general, $\mathcal{C} \cap V^n$ is not $\mathbb{F}_{q^m}$−linear, but it is $\mathbb{F}_q$−linear and also linear over some intermediate extension depending on V. This code also corresponds by projection to the Subgroup Subcode [19] on the alphabet $\mathbb{F}_q^s$.

In the following, we introduce a generalization of Definition 4 by allowing the choice of different vector subspaces for each coordinate.

Definition 5 Let $\mathcal{C}$ be a $[n, k, d = d_R(\mathcal{C})]_{q^m}$−code, and $V_1, \ldots, V_n$ a series of n $\mathbb{F}_q$−vector subspaces of $\mathbb{F}_{q^m}$ with dimensions respectively $s_1, s_2, \ldots, s_n$. Let denotes $\prod_{i=1}^{n} V_i$ the Cartesian product of $V_1, \ldots, V_n$. The Rank Generalized Subspace Subcode of $\mathcal{C}$ with respect to $\prod_i = 1^n V_i$ is the $\mathbb{F}_q$−vector space $\mathcal{C} \cap \prod_i = 1^n V_i$.

Remark that all the V_i's are equal in Definition 4, contrary to what we have in the above Definition 5.

Let $B = \{b_1, b_2, \ldots, b_m\}$ be a basis of $\mathbb{F}_{q^m}$ as a $\mathbb{F}_q$−vector space. We consider the map $\phi_B : \mathbb{F}_{q^m} \longrightarrow \mathbb{F}_q^m$, mapping each element of $\mathbb{F}_{q^m}$ to its $\mathbb{F}_q$−coordinates in the basis B. That is to say, for any $x = \sum_{i=1}^{m} x_i b_i \in \mathbb{F}_{q^m}$ (with $x_i \in \mathbb{F}_q$), $\phi_B(x) = (x_1, x_2, \ldots, x_m)$. Given n $\mathbb{F}_q$−bases $B_1, \ldots, B_n$ of $\mathbb{F}_{q^m}$, this map can be extended to $\mathbb{F}_{q^m}^n$ by

$$\begin{aligned}\phi_{(B_i)_i} : \mathbb{F}_{q^m}^n &\longrightarrow \mathbb{F}_q^{mn}\\ (c_1, c_2, \ldots, c_n) &\longmapsto (\phi_{B_1}(c_1), \phi_{B_2}(c_2), \ldots, \phi_{B_n}(c_n)),\end{aligned}$$

In the case $B_1 = \cdots = B_n = B$, we will simply use ϕ_B instead of $\phi_{(B_i)_i}$.

$\phi_{(B_i)_i}$ is called the expansion function and, applying it to all codewords of a $[n, k]$−code $\mathcal{C}$, this gives a new linear code of length nm called q−ary image of $\mathcal{C}$ and denoted by $\mathcal{C}(B_i)_i$ or simply $\mathcal{C}B$ if $B_1 = \cdots = B_n = B$. More formally, we have the following definition.

Definition 6 Let $\mathcal{C}$ be a linear code of length n and dimension k over $\mathbb{F}_{q^m}$. The $q-$ary image $\mathcal{C}(B_i)_i$ of $\mathcal{C}$ with respect to the n bases $B_1, \ldots, B_n$ of $\mathbb{F}_{q^m}$ is the image of $\mathcal{C}$ by the $\mathbb{F}_q-$linear map $\phi_{(B_i)_i}$:

$$\mathcal{C}(B_i)_i = \phi_{(B_i)_i}(\mathcal{C})$$

Let $\mathcal{C}$ be an $[n, k]$ linear code over $\mathbb{F}_{q^m}$. In the following proposition, we describe how to explicitly compute a generator matrix and a parity-check matrix for the q-ary image $\mathcal{C}(B_i)_i$ of $\mathcal{C}$, where each coordinate expansion is done with respect to a possibly distinct basis B_i of $\mathbb{F}_{q^m}$.

Proposition 3 *Let $\mathcal{C}$ be a $[n, k]-$code over $\mathbb{F}_{q^m}$ and $B = \{b_1, b_2, \ldots, b_m\}$ be a basis of $\mathbb{F}_{q^m}$.*

1. *If $\boldsymbol{G} = \begin{pmatrix} \mathbf{g}_1 \\ \mathbf{g}_2 \\ \vdots \\ \mathbf{g}_k \end{pmatrix} \in \mathbb{F}_{q^m}^{k\times n}$ is a generator matrix of $\mathcal{C}$ then, the matrix $\hat{\boldsymbol{G}}^B_{(B_j)_j}$ defined by*

$$\hat{\boldsymbol{G}}^B_{(B_j)_j} = \begin{pmatrix} \phi_{(B_j)_j}(b_1\mathbf{g}_i) \\ \phi_{(B_j)_j}(b_2\mathbf{g}_i) \\ \vdots \\ \phi_{(B_j)_j}(b_m\mathbf{g}_i) \end{pmatrix}_{1\leq i\leq k} \in \mathbb{F}_q^{mk\times mn} \tag{1}$$

is a generator matrix of the $q-$ary image $\mathcal{C}(B_i)_i$ of $\mathcal{C}$ with respect to the n bases $B_1, \ldots, B_n$ of $\mathbb{F}_{q^m}$.

2. *If $\boldsymbol{H} = \begin{pmatrix} \mathbf{h}_1 \\ \mathbf{h}_2 \\ \vdots \\ \mathbf{h}_n \end{pmatrix}^T \in \mathbb{F}_{q^m}^{(n-k)\times n}$ is a parity-check[2] matrix of $\mathcal{C}$, then the matrix $\hat{\boldsymbol{H}}^B_{(B_j)_j}$ defined by*

$$\hat{\boldsymbol{H}}^B_{(B_j)_j} = \begin{pmatrix} \phi_{(B_j)_j}(b_1\mathbf{h}_i) \\ \phi_{(B_j)_j}(b_2\mathbf{h}_i) \\ \vdots \\ \phi_{(B_j)_j}(b_m\mathbf{h}_i) \end{pmatrix}^T_{1\leq i\leq n} \in \mathbb{F}_q^{m(n-k)\times mn} \tag{2}$$

is a parity-check matrix of the $q-$ary image $\mathcal{C}(B_i)_i$ of $\mathcal{C}$ with respect to the $n-k$ bases $B_1, \ldots, B_{n-k}$ of $\mathbb{F}_{q^m}$.

[2] Here, the $\mathbf{h}_i$ are the column vectors of the matrix $\boldsymbol{H}$.

Proof For the first point, let us consider for any line $i \in \{1, \cdots, k\}$ of $\boldsymbol{G}$, the map

$$\begin{aligned} f_{\mathbf{g}_i} : \mathbb{F}_{q^m} &\longrightarrow \mathbb{F}_{q^m}^n \\ x &\longmapsto f_{\mathbf{g}_i}(x) = x\mathbf{g}_i. \end{aligned}$$

Since ϕ_B is an $\mathbb{F}_q-$isomorphism from $\mathbb{F}_{q^m}$ to $\mathbb{F}_q^m$, we consider the $\mathbb{F}_q-$linear map $\theta_{\mathbf{g}_i} = \phi_{(B_j)_j} \circ f_{\mathbf{g}_i} \circ \phi_B^{-1}$ from $\mathbb{F}_q^m$ to $\mathbb{F}_q^{mn}$. Let $B^* = (b_1^*, \ldots, b_m^*)$ be a basis of $\mathbb{F}_q^m$ such that for any $i \in \{1, \cdots, m\}$, $b_i^* = \phi_B(b_i)$. As the encoding is considered as a vector-matrix product, the matrix of $\theta_{\mathbf{g}_i}$ in the basis B^* is given by a matrix $\boldsymbol{M}_{\mathbf{g}_i} \in \mathbb{F}_q^{m\times mn}$ such that:

$$\boldsymbol{M}_{\mathbf{g}_i} = \begin{pmatrix} \theta_{\mathbf{g}_i}(b_1^*) \\ \theta_{\mathbf{g}_i}(b_2^*) \\ \vdots \\ \theta_{\mathbf{g}_i}(b_m^*) \end{pmatrix} = \begin{pmatrix} \phi_{(B_j)_j} \circ f_{\mathbf{g}_i} \circ \phi_B^{-1}(b_1^*) \\ \phi_{(B_j)_j} \circ f_{\mathbf{g}_i} \circ \phi_B^{-1}(b_2^*) \\ \vdots \\ \phi_{(B_j)_j} \circ f_{\mathbf{g}_i} \circ \phi_B^{-1}(b_m^*) \end{pmatrix} = \begin{pmatrix} \phi_{(B_j)_j}(b_1\mathbf{g}_i) \\ \phi_{(B_j)_j}(b_2\mathbf{g}_i) \\ \vdots \\ \phi_{(B_j)_j}(b_m\mathbf{g}_i) \end{pmatrix}$$

To finish, we have by definition

$$\begin{aligned} \phi_{(B_j)_j}(\mathcal{C}) &= \{\phi_{(B_j)_j}(\mathbf{m}\boldsymbol{G}), \mathbf{m} \in \mathbb{F}_{q^m}^k\} \\ &= \{\phi_{(B_j)_j}(\textstyle\sum_{i=1}^k m_i\mathbf{g}_i), \mathbf{m} = (m_1, \cdots, m_k) \in \mathbb{F}_{q^m}^k\} \\ &= \{\sum_{i=1}^k \phi_{(B_j)_j} \circ f_{\mathbf{g}_i}(m_i), \mathbf{m} = (m_1, \cdots, m_k) \in \mathbb{F}_{q^m}^k\} \\ &= \{\sum_{i=1}^k \theta_{\mathbf{g}_i} \circ \phi_B(m_i), \mathbf{m} = (m_1, \cdots, m_k) \in \mathbb{F}_{q^m}^k\} \\ &= \{\sum_{i=1}^k \theta_{\mathbf{g}_i}(\mathbf{m}_i'), \mathbf{m}' = (\mathbf{m}_1', \cdots, \mathbf{m}_k') \in (\mathbb{F}_q^m)^k\} \\ &= \{\sum_{i=1}^k \mathbf{m}_i'\boldsymbol{M}_{\mathbf{g}_i}, \mathbf{m}' = (\mathbf{m}_1', \cdots, \mathbf{m}_k') \in (\mathbb{F}_q^m)^k\} \\ &= \{\mathbf{m}' \cdot \begin{pmatrix} \boldsymbol{M}_{\mathbf{g}_1} \\ \boldsymbol{M}_{\mathbf{g}_2} \\ \vdots \\ \boldsymbol{M}_{\mathbf{g}_k} \end{pmatrix}, \quad \mathbf{m}' \in \mathbb{F}_q^{km}\} \\ &= < \begin{pmatrix} \boldsymbol{M}_{\mathbf{g}_1} \\ \boldsymbol{M}_{\mathbf{g}_2} \\ \vdots \\ \boldsymbol{M}_{\mathbf{g}_k} \end{pmatrix} >_{\mathbb{F}_q} \end{aligned}$$

The proof of the second point of the proposition is similar. Indeed, by considering only the $n-k$ bases $B_1, \ldots, B_{n-k}$ of $\mathbb{F}_{q^m}$, let us apply $\phi_{(B_j)_j}$ on a syndrome $\mathbf{s}$ given by $\mathbf{s}^T = \mathbf{y} \cdot H^T$. We have

$$\begin{aligned}
\phi_{(B_j)_j}(\mathbf{s}^T) &= \phi_{(B_j)_j}(\mathbf{y}\cdot H^T)\\
&= \sum_{i=1}^{n} \phi_{(B_j)_j} \circ f_{\mathbf{h}_i}(y_i)\\
&= \phi_B(\mathbf{y}) \cdot \begin{pmatrix} \boldsymbol{M}_{\mathbf{h}_1} \\ \boldsymbol{M}_{\mathbf{h}_2} \\ \vdots \\ \boldsymbol{M}_{\mathbf{h}_n} \end{pmatrix} = \phi_B(\mathbf{y}) \cdot \begin{pmatrix} \phi_{(B_j)_j}(b_1\mathbf{h}_i) \\ \phi_{(B_j)_j}(b_2\mathbf{h}_i) \\ \vdots \\ \phi_{(B_j)_j}(b_m\mathbf{h}_i) \end{pmatrix}_{1\le i\le n}
\end{aligned}$$

One can remark that this proposition is a general case of [32, Theorem 1], which can be obtained from the above proposition by taking $B_1 = B_2 = \cdots = B_n$. Furthermore, by using the notations in the above proof, one can remark that for any $i \in \{1, \cdots, k\}$, the map $f_{\mathbf{g}_i}$ can be decomposed as $f_{\mathbf{g}_i} = (f_{g_{i1}}, f_{g_{i2}}, \cdots, f_{g_{in}})$. So, the matrix $\hat{\boldsymbol{G}}^{B}_{(B_j)_j}$ can be written as follows:

$$\hat{\boldsymbol{G}}^{B}_{(B_j)_j} = (\boldsymbol{M}_{g_{ij}})_{1\le i\le k, 1\le j\le n} = \begin{pmatrix} \boldsymbol{M}_{g_{11}} & \cdots & \boldsymbol{M}_{g_{1n}} \\ \boldsymbol{M}_{g_{21}} & \cdots & \boldsymbol{M}_{g_{2n}} \\ \vdots & \ddots & \vdots \\ \boldsymbol{M}_{g_{k1}} & \cdots & \boldsymbol{M}_{g_{kn}} \end{pmatrix} \in (\mathbb{F}_q^{m\times m})^{k\times n}$$

where the matrix $\boldsymbol{M}^{T}_{g_{ij}} = \mathcal{M}_{B,B_j}(\theta_{g_{ij}})$ is the matrix of the linear map $\theta_{g_{ij}}$ from the basis B to the basis B_j, for $i \in \{1, \cdots, k\}$ and $j \in \{1, \cdots, n\}$. Another remark is that, for a different basis B' we get another generator matrix $\hat{\boldsymbol{G}}^{B'}_{(B_j)_j}$ of the same code, and this is the same for $\hat{\boldsymbol{H}}^{B'}_{(B_j)_j}$. To simplify things in the following, we will sometimes write $\hat{\boldsymbol{G}}_{(B_j)_j}$ instead of $\hat{\boldsymbol{G}}^{B}_{(B_j)_j}$ and if $B_1 = B_2 = \cdots = B_n = B$, we define $\hat{\boldsymbol{G}}_B := \hat{\boldsymbol{G}}^{B}_{(B_j)_j}$.

Example 1 Let $n = 7$, $m = 3$, and let α be a root of the irreducible polynomial $x^3 + x + 1$. Let's consider the $[n, 3]-$code $\mathcal{C}$ generated by the following generator matrix:

$$\boldsymbol{G} = \begin{pmatrix} 1\,0\,0\,\alpha^3\ \alpha\ 1\ \alpha^3 \\ 0\,1\,0\,\alpha^6\ \alpha^6\ 1\ \alpha^2 \\ 0\,0\,1\,\alpha^5\ \alpha^4\ 1\ \alpha^4 \end{pmatrix}$$

For $\mathcal{B} = \{1 = (100), \alpha = (010), \alpha^2 = (001)\} = B_1 = \cdots B_7$, a generator matrix of the binary image $\mathcal{C}B$ of $\mathcal{C}$ is:

$$\hat{\boldsymbol{G}}_B = \left(\begin{array}{ccc|ccc|ccc|ccc|ccc|ccc|ccc}
1&0&0&0&0&0&0&0&0&1&1&0&0&1&0&1&0&0&1&1&0\\
0&1&0&0&0&0&0&0&0&0&1&1&0&0&1&0&1&0&0&1&1\\
0&0&1&0&0&0&0&0&0&1&1&1&1&1&0&0&0&1&1&1&1\\
\hline
0&0&0&1&0&0&0&0&0&1&0&1&1&0&1&1&0&0&0&0&1\\
0&0&0&0&1&0&0&0&0&1&0&0&1&0&0&0&1&0&1&1&0\\
0&0&0&0&0&1&0&0&0&0&1&0&0&1&0&0&0&1&0&1&1\\
\hline
0&0&0&0&0&0&1&0&0&1&1&1&0&1&1&1&0&0&0&1&1\\
0&0&0&0&0&0&0&1&0&1&0&1&1&1&1&0&1&0&1&1&1\\
0&0&0&0&0&0&0&0&1&1&0&0&1&0&1&0&0&1&1&0&1
\end{array}\right)$$

A parity-check matrix of $\mathcal{C}B$ is, by using the fact that $\hat{\boldsymbol{G}}_B \cdot (\hat{\boldsymbol{H}}_B)^T = 0$:

$$\hat{\boldsymbol{H}}_B = \left(\begin{array}{ccc|ccc|ccc|ccc|ccc|ccc|ccc}
1&0&1&1&1&0&1&1&1&1&0&0&0&0&0&0&0&0&0&0&0\\
1&1&1&0&0&1&1&0&0&0&1&0&0&0&0&0&0&0&0&0&0\\
0&1&1&1&0&0&1&1&0&0&0&1&0&0&0&0&0&0&0&0&0\\
\hline
0&0&1&1&1&0&0&1&1&0&0&0&1&0&0&0&0&0&0&0&0\\
1&0&1&0&0&1&1&1&0&0&0&0&0&1&0&0&0&0&0&0&0\\
0&1&0&1&0&0&1&1&1&0&0&0&0&0&1&0&0&0&0&0&0\\
\hline
1&0&0&1&0&0&1&0&0&0&0&0&0&0&0&1&0&0&0&0&0\\
0&1&0&0&1&0&0&1&0&0&0&0&0&0&0&0&1&0&0&0&0\\
0&0&1&0&0&1&0&0&1&0&0&0&0&0&0&0&0&1&0&0&0\\
\hline
1&0&1&0&1&0&0&1&1&0&0&0&0&0&0&0&0&0&1&0&0\\
1&1&1&0&1&1&1&1&0&0&0&0&0&0&0&0&0&0&0&1&0\\
0&1&1&1&0&1&1&1&1&0&0&0&0&0&0&0&0&0&0&0&1
\end{array}\right)$$

To construct a subcode restricted to subspaces, the general approach begins by computing the $q-$ary image of the code via a fixed basis of $\mathbb{F}_{q^m}$. Then, for each coordinate block of size m, a suitable change of basis is applied. After this transformation, the code is punctured on selected positions to eliminate undesired coordinates, and taking the dual of the resulting code yields the desired subcode. In what follows, we present several results underlying this process.

In the sequel, given a set of positions $U \subseteq \{1, 2, \ldots, nm\}$, we denote its complement by $\bar{U} := \{1, 2, \ldots, nm\} \setminus U$. The next proposition formalizes this construction: it shows that shortening the q-ary image of a code $\mathcal{C}$ on the positions $\bar{U}$ yields the q-ary image of a subspace subcode $\mathcal{C} \cap V^n$, where $V \subset \mathbb{F}_{q^m}$ is an appropriately chosen $\mathbb{F}_q$-subspace. Since $\mathcal{C} \cap V^n \subset V^n$, each codeword can be expressed using a basis B_s of V. We will abusively denote by $(\mathcal{C} \cap V^n)B_s$ the q-ary image obtained via coordinate-wise expansion with respect to B_s.

Proposition 4 *Let $\mathcal{C} \subseteq \mathbb{F}_{q^m}^n$ be a linear code of length n, $B = \{b_1, b_2, \ldots, b_m\}$ an $\mathbb{F}_q$-basis of $\mathbb{F}_{q^m}$, and let $B_s = \{b_1, \ldots, b_s\}$ with $V = \langle B_s \rangle \subset \mathbb{F}_{q^m}$ an s-dimensional $\mathbb{F}_q$-subspace. Define the set $U = u_1 \cup \ldots \cup u_n \subseteq \{1, 2, \ldots, nm\}$, where for each $1 \leq j \leq n$, we let $u_j = \{(j-1)m + 1, \ldots, (j-1)m + s\}$. Then the shortened code $(\mathcal{C}B)\bar{U}$ is equal to the q-ary image of the subspace subcode $\mathcal{C} \cap V^n$ with respect to the basis B_s:*

$$(\mathcal{C} \cap V^n)B_s = (\mathcal{C}B)\bar{U}$$

Proof Let $\mathbf{y} = (y_{1,1}, \ldots, y_{s,1}, \ldots, y_{1,j}, \ldots, y_{s,j}, \ldots, y_{1,n}, \ldots, y_{s,n}) \in (\mathcal{C}B)\bar{U}$.

Then there exists $\mathbf{c} \in \phi_B(\mathcal{C}) = \mathcal{C}B$ such that $\mathbf{y}$ is obtained by puncturing $\mathbf{c}$ at the positions in $\bar{U}$, and $\mathbf{c}_i = 0$ for all $i \in \bar{U}$.

Since $V = \phi_B^{-1}(\{(x_1, \ldots, x_s, 0, \ldots, 0) \mid x_1, \ldots, x_s \in \mathbb{F}_q\})$, it follows that $\phi_B^{-1}(\mathbf{c}) \in \mathcal{C} \cap V^n$, and thus $\mathbf{c} \in \phi_B(\mathcal{C} \cap V^n)$. Hence,

$$\mathbf{y} \in ((\mathcal{C} \cap V^n)B)\bar{U} = (\mathcal{C} \cap V^n)B_s.$$

Conversely, it is immediate that $(\mathcal{C} \cap V^n)B_s \subseteq (\mathcal{C}B)\bar{U}$, so we conclude:

$$(\mathcal{C} \cap V^n)B_s = (\mathcal{C}B)\bar{U}$$

Thanks to Proposition 4, the construction of a subspace subcode $\mathcal{C} \cap V^n$ can, in practice, be reduced to computing the q-ary image of $\mathcal{C}$ and then shortening it at appropriately chosen positions. To do so, we may use Theorem 1, $((\mathcal{C}B)\bar{U})^{\perp} = ((\mathcal{C}B)^{\perp})\bar{U}$ and compute the punctured code that is easier to do.

Example 2 Using the same code defined in Example 1, we consider $s = 1$ and $V = \mathbb{F}_2 = \langle 1 \rangle_{\mathbb{F}_2}$ to mean $B_s = \{1\}$. By definition, this yields the index set

$$U = u_1 \cup \cdots \cup u_7 = \{1, 4, 7, 10, 13, 16, 19\},$$

corresponding to the first coordinate of each block of size $m = 3$. Thanks to Theorem 1, the dual of the shortened code $(\mathcal{C}B)\bar{U}$ is generated by the matrix $(\hat{\boldsymbol{H}}_B)\bar{U}$ obtained from the full parity-check matrix $\hat{\boldsymbol{H}}_B$ by retaining only the columns indexed by U. That is to say

$$(\hat{\boldsymbol{H}}_B)\bar{U} = \begin{pmatrix} 1\,1\,1\,1\,0\,0\,0 \\ 1\,0\,1\,0\,0\,0\,0 \\ 0\,1\,1\,0\,0\,0\,0 \\ \hline 0\,1\,0\,0\,1\,0\,0 \\ 1\,0\,1\,0\,0\,0\,0 \\ 0\,1\,1\,0\,0\,0\,0 \\ \hline 1\,1\,1\,0\,0\,1\,0 \\ 0\,0\,0\,0\,0\,0\,0 \\ 0\,0\,0\,0\,0\,0\,0 \\ \hline 1\,0\,0\,0\,0\,0\,1 \\ 1\,0\,1\,0\,0\,0\,0 \\ 0\,1\,1\,0\,0\,0\,0 \end{pmatrix}$$

Thus, we obtain that $((\mathcal{C}B)\bar{U})^{\perp} = (\mathcal{C} \cap \mathbb{F}_2^7)^{\perp}$ is generated by

$$\boldsymbol{H}_{\bar{U}} = \begin{pmatrix} 1&1&1&1&0&0&0 \\ 1&0&1&0&0&0&0 \\ 0&1&1&0&0&0&0 \\ 0&1&0&0&1&0&0 \\ 1&1&1&0&0&1&0 \\ 1&0&0&0&0&0&1 \end{pmatrix}$$

Note that one can easily compute a generator matrix $\boldsymbol{G}_{\bar{U}}$ of the subspace subcode $\mathcal{C} \cap \mathbb{F}_2^7$ from $\boldsymbol{H}_{\bar{U}}$.

We now recall a fundamental result describing how a coordinate-wise change of basis in $\mathbb{F}_{q^m}^n$ impacts the q-ary image of a code, as well as its associated generator and parity-check matrices.

Theorem 2 ([5], Lemma 34) *Let $\mathcal{C}$ be a linear code of length n over $\mathbb{F}_{q^m}$, $B = \{b_1, b_2, \ldots, b_m\}$ be a basis of $\mathbb{F}_{q^m}$. Let $(\boldsymbol{Q}_j)_j \in (GL_m(\mathbb{F}_q))^n$. The following equalities hold.*

$$\mathcal{C}(\boldsymbol{Q}_j^{-1}B)_j = \mathcal{C}B \cdot \begin{pmatrix} \boldsymbol{Q}_1 & & \\ & \ddots & \\ & & \boldsymbol{Q}_n \end{pmatrix},$$

$$\hat{\boldsymbol{G}}^B_{(\boldsymbol{Q}_j^{-1}B)_j} = \hat{\boldsymbol{G}}_B \cdot \begin{pmatrix} \boldsymbol{Q}_1 & & \\ & \ddots & \\ & & \boldsymbol{Q}_n \end{pmatrix}, \textit{ and}$$

$$\hat{\boldsymbol{H}}^B_{(\boldsymbol{Q}_j^{-1}B)_j} = \hat{\boldsymbol{H}}_B \cdot \begin{pmatrix} (\boldsymbol{Q}_1^{-1})^T & & \\ & \ddots & \\ & & (\boldsymbol{Q}_n^{-1})^T \end{pmatrix}$$

In practice, Theorem 2 implies that changing the basis at each coordinate position corresponds to right-multiplication by a block-diagonal matrix on the q-ary image. This technique offers several important benefits. Firstly, it allows working from a global basis (e.g., the canonical basis), simplifying the expansion process via standard coordinate maps. Secondly, the change-of-basis matrices can be structured so that their first s_i columns span the subspace V_i, and the remaining columns form a complement, enabling efficient coordinate elimination through puncturing. During decoding, this layout also helps easily identify the eliminated positions, facilitating reliable error correction in the subspace subcode. These features make this approach well suited for the construction and decoding of generalized subspace subcodes. Consequently, we deduce the following corollary.

Corollary 1 *Let $\mathcal{C}$ be a linear code of length n over $\mathbb{F}_{q^m}$, $B = \{b_1, b_2, \ldots, b_m\}$ be a basis of $\mathbb{F}_{q^m}$ and $V =< d_1, \ldots, d_s >_{\mathbb{F}_q}$ a $s-$dimensional subspaces of $\mathbb{F}_{q^m}$. Given any completed basis $D = \{d_1, d_2, \ldots, d_s, \ldots, d_m\}$ from $D_s = \{d_1, d_2, \ldots, d_s\}$, and $\boldsymbol{Q}$ the change-of-basis matrix from the basis D to the basis B, we have:*

$$(\mathcal{C}D)\bar{U} = (\mathcal{C} \cap V^n)B \cdot \begin{pmatrix} \boldsymbol{Q} & & \\ & \ddots & \\ & & \boldsymbol{Q} \end{pmatrix} \bar{U}$$

Where $U = u_1 \cup \ldots \cup u_n \subseteq \{1, 2, \ldots, nm\}$ *with* $u_j = \{(j-1)m+1, \ldots, (j-1)m+s\}$ *for* $1 \leq j \leq n$.

Before giving the proof, we note that the change-of-basis matrices are defined with the convention that message encoding is performed via vector-matrix multiplication (i.e., messages are treated as row vectors). Consequently, the change of basis is applied row-wise. When we represent a basis $B = \{b_1, b_2, \ldots, b_m\}$ as a matrix, it simply corresponds to the change-of-basis matrix from the canonical basis $K = \{1, \alpha, \ldots, \alpha^{m-1}\}$ to the chosen basis: $B = \begin{pmatrix} \phi_K(b_1) \\ \phi_K(b_2) \\ \vdots \\ \phi_K(b_m) \end{pmatrix} \in \mathbb{F}_q^{m \times m}$

Proof According to Proposition 4, we have

$$(\mathcal{C}D)\bar{U} = (\mathcal{C} \cap V^n)D_s = ((\mathcal{C} \cap V^n)D)\bar{U}.$$

As $\boldsymbol{Q}D = B$, then

$$(\mathcal{C} \cap V^n)D_s = \left((\mathcal{C} \cap V^n)Q^{-1}B\right)\bar{U} = (\mathcal{C} \cap V^n)B \cdot \begin{pmatrix} \boldsymbol{Q} & & \\ & \ddots & \\ & & \boldsymbol{Q} \end{pmatrix} \bar{U}.$$

We now extend the previous result to the case where each coordinate of the code is restricted to a different subspace $V_j \subseteq \mathbb{F}_{q^m}$. This leads to the following expression for the q-ary image of a generalized subspace subcode.

Corollary 2 *Let* $\mathcal{C}$ *be a linear code of length* n *over* $\mathbb{F}_{q^m}$ *and* B *be a* $\mathbb{F}_q$*−basis of* $\mathbb{F}_{q^m}$. *For any* $j \in \{1, \cdots, m\}$, *let* $V_j =< d_{1,j}, \ldots, d_{s_j,j} >$ *be a* s_j*−dimensional* $\mathbb{F}_q$*−subspaces of* $\mathbb{F}_{q^m}$, $D_j = \{d_{1,j}, \ldots, d_{s_j,j}, \ldots, d_{m,j}\}$ *be a completed basis from* $D_{s_j} = \{d_{1,j}, \ldots, d_{s_j,j}\}$, *and* $\boldsymbol{Q}_j$ *the change-of-basis matrix from the basis* D_j *to the basis* D. *By setting* $W = \prod_{i=1}^{n} V_i$, *we have*

$$(\mathcal{C}(D_j)_j)\bar{U} = ((\mathcal{C} \cap W)(\boldsymbol{Q}_j^{-1}B)_j)\bar{U} = (\mathcal{C} \cap W)B \cdot \begin{pmatrix} \boldsymbol{Q}_1 & & \\ & \ddots & \\ & & \boldsymbol{Q}_n \end{pmatrix} \bar{U}$$

Where $U = u_1 \cup \ldots \cup u_n \subseteq \{1, 2, \ldots, nm\}$ *with* $u_j = \{(j-1)m+1, \ldots, (j-1)m+s_j\}$ *for* $1 \leq j \leq n$.

Thanks to the duality between shortening and puncturing (Theorem 1), a generator matrix of the generalized subspace subcode $(\mathcal{C} \cap W)(\boldsymbol{Q}_j^{-1}B)_j$ can be obtained by computing the kernel of the punctured parity-check matrix $(\hat{\boldsymbol{H}}_{(\boldsymbol{Q}_j^{-1}B)_j})\bar{U}$.

Example 1 Using the same code defined in Example 1, let us define $W = V_1 \times V_2 \times V_1 \times V_3 \times V_1 \times V_2 \times V_1$, with $V_1 = \langle 1, \alpha \rangle$, $V_2 = \langle 1, \alpha^2 \rangle$, and $V_3 = \langle \alpha, \alpha^2 \rangle$. We consider $D_1 = D_3 = D_5 = D_7 = \mathcal{B} = \{1, \alpha, \alpha^2\}$, $D_2 = D_6 = \{1, \alpha^2, \alpha\}$, and $D_4 = \{\alpha, \alpha^2, 1\}$. It is obvious that all the s_j have the same value, which is 2 and $U = u_1 \cup \cdots \cup u_7 = \{1, 2, 4, 5, 7, 8, 10, 11, 13, 14, 16, 17, 19, 20\}$. The different change-of-basis matrices from D_j to the bases B $(j = 1, \ldots, 7)$ are :

$$\boldsymbol{Q}_1 = \boldsymbol{Q}_3 = \boldsymbol{Q}_5 = \boldsymbol{Q}_7 = \boldsymbol{I}_3, \ \boldsymbol{Q}_2 = \boldsymbol{Q}_6 = \begin{pmatrix} 1 & 0 & 0 \\ 0 & 0 & 1 \\ 0 & 1 & 0 \end{pmatrix} \text{ and } \boldsymbol{Q}_4 = \begin{pmatrix} 0 & 1 & 0 \\ 0 & 0 & 1 \\ 1 & 0 & 0 \end{pmatrix}$$

Let $\boldsymbol{D} = \mathrm{Diag}(\boldsymbol{Q}_1^T, \boldsymbol{Q}_2^T, \boldsymbol{Q}_3^T, \boldsymbol{Q}_4^T, \boldsymbol{Q}_5^T, \boldsymbol{Q}_6^T, \boldsymbol{Q}_7^T)\bar{U}$. Then an extended parity-check matrix of the subcode $\mathcal{C} \cap W$ is given by $H_U = \hat{\boldsymbol{H}}_B \cdot D$.

Hence, H_U is obtained by deleting columns $3, 5, 9, 10, 15, 17, 21$ from $\hat{\boldsymbol{H}}_B$, resulting in:

$$H_U = \left(\begin{array}{cc|cc|cc|cc|cc|cc|cc}
1 & 0 & 1 & 0 & 1 & 1 & 0 & 0 & 0 & 0 & 0 & 0 & 0 & 0 \\
1 & 1 & 0 & 1 & 1 & 0 & 1 & 0 & 0 & 0 & 0 & 0 & 0 & 0 \\
0 & 1 & 1 & 0 & 1 & 1 & 0 & 1 & 0 & 0 & 0 & 0 & 0 & 0 \\
\hline
0 & 0 & 1 & 0 & 0 & 1 & 0 & 0 & 1 & 0 & 0 & 0 & 0 & 0 \\
1 & 0 & 0 & 1 & 1 & 1 & 0 & 0 & 0 & 1 & 0 & 0 & 0 & 0 \\
0 & 1 & 1 & 0 & 1 & 1 & 0 & 0 & 0 & 0 & 0 & 0 & 0 & 0 \\
\hline
1 & 0 & 1 & 0 & 1 & 0 & 0 & 0 & 0 & 0 & 1 & 0 & 0 & 0 \\
0 & 1 & 0 & 0 & 0 & 1 & 0 & 0 & 0 & 0 & 0 & 0 & 0 & 0 \\
0 & 0 & 0 & 1 & 0 & 0 & 0 & 0 & 0 & 0 & 0 & 1 & 0 & 0 \\
\hline
1 & 0 & 0 & 0 & 0 & 1 & 0 & 0 & 0 & 0 & 0 & 0 & 1 & 0 \\
1 & 1 & 0 & 1 & 1 & 1 & 0 & 0 & 0 & 0 & 0 & 0 & 0 & 1 \\
0 & 1 & 1 & 1 & 1 & 1 & 0 & 0 & 0 & 0 & 0 & 0 & 0 & 0
\end{array}\right)$$

After row operations, we obtain an extended generator matrix of $\mathcal{C} \cap W$ in a systematic form with respect to the bases D_j:

$$G_U = \left(\begin{array}{cc|cc|cc|cc|cc|cc|cc}
1 & 0 & 1 & 1 & 0 & 0 & 0 & 1 & 1 & 0 & 0 & 1 & 1 & 1 \\
0 & 1 & 0 & 1 & 1 & 1 & 1 & 1 & 1 & 1 & 1 & 1 & 1 & 1
\end{array}\right)$$

As a binary code, $\mathcal{C}$ has length $n = 14$ and dimension $k = 2$. As a block code, it has length $n = 7$ and dimension $k = 2$.

Let's now establish a lower bound on the cardinality of a generalized subspace subcode in the rank-metric, expressed in terms of the dimensions of the underlying subspaces.

Proposition 5 *Let $\mathcal{C}$ be a $[n, k, d = d_R(\mathcal{C})]_{q^m}$-code, and let $V_1, \ldots, V_n$ be $\mathbb{F}_q$-subspaces of $\mathbb{F}_{q^m}$ of respective dimensions $s_1, \ldots, s_n \leq m$. Define the ambient subspace $W := \prod_{i=1}^n V_i \subseteq \mathbb{F}_{q^m}^n$. Then the following inequality holds:*

$$q^{\sum_{i=1}^n s_i - m(n-k)} \leq |\mathcal{C} \cap W|.$$

Proof Without loss of generality, assume that $s_1 \leq s_2 \leq \cdots \leq s_n$. By permuting the coordinates of $\mathcal{C}$, which preserves its code parameters and subcode dimension (up to isometry), we may assume that $V_1 \subseteq V_2 \subseteq \cdots \subseteq V_n$.

Let $\{\beta_1, \ldots, \beta_{s_n}\}$ be a common $\mathbb{F}_q$-basis such that $V_i = \langle \beta_1, \ldots, \beta_{s_i} \rangle$. Then any vector $\mathbf{c} = (c_1, \ldots, c_n) \in W$ can be written as:

$$c_i = \sum_{t=1}^{s_i} u_{t,i} \beta_t = \sum_{t=1}^{s_n} u_{t,i} \beta_t, \quad \text{where } u_{t,i} = 0 \text{ for } t > s_i.$$

Hence,

$$\mathbf{c} = (\beta_1, \ldots, \beta_{s_n}) \cdot \boldsymbol{U}, \quad \text{where } \boldsymbol{U} \in \mathbb{F}_q^{s_n \times n}.$$

Let $\boldsymbol{H} \in \mathbb{F}_{q^m}^{(n-k)\times n}$ be a parity-check matrix of $\mathcal{C}$. Then $\mathbf{c} \in \mathcal{C}$ if and only if :

$$(\beta_1, \ldots, \beta_{s_n}) \cdot \boldsymbol{U} \cdot \boldsymbol{H}^T = 0.$$

Expanding each product $\beta_i h_{j,t}$ over a fixed $\mathbb{F}_q$-basis $\{\gamma_1, \ldots, \gamma_m\}$ of $\mathbb{F}_{q^m}$, we write:

$$\beta_i h_{j,t} = \sum_{\ell=1}^{m} \delta_{i,t}^{(j,\ell)} \gamma_\ell, \quad \delta_{i,t}^{(j,\ell)} \in \mathbb{F}_q.$$

So the condition above leads to the following linear system over $\mathbb{F}_q$ in the unknowns $u_{t,i}$:

$$\sum_{i=1}^{s_n} \sum_{t=1}^{n} \delta_{i,t}^{(j,\ell)} u_{i,t} = 0 \quad \text{for all } j = 1, \ldots, n-k, \ \ell = 1, \ldots, m.$$

This gives a system of $m(n-k)$ linear equations in $\sum_{i=1}^n s_i$ unknowns. Hence, the space of solutions has dimension at least $\sum_{i=1}^n s_i - m(n-k)$, which yields:

$$|\mathcal{C} \cap W| \geq q^{\sum_{i=1}^n s_i - m(n-k)}.$$

Notice that taking $s_1 = \ldots = s_n$ leads to the result obtained by Gabidulin and Loidreau in [11] as a special case.

Corollary 3 [[11], Proposition 1] *Let $\mathcal{C}$ be a $[n, k, d = d_R(\mathcal{C})]_{q^m}$-code and let $V \subseteq \mathbb{F}_{q^m}$ be an $\mathbb{F}_q$-subspace of dimension $s \leq m$. Then,*

$$q^{ns-m(n-k)} \leq |\mathcal{C} \cap V^n|.$$

To summarize, we have established an explicit representation of generalized subspace subcodes using expansion maps, change-of-basis matrices, and shortening operations. These tools pave the way for efficient construction and analysis of subcodes with tailored structural properties.

4 Generalized Subspace Subcodes of Gabidulin Codes

4.1 Gabidulin Codes

Gabidulin codes [9] are a class of maximum rank distance (MRD) codes constructed using q-polynomials, also known as linearized polynomials, first studied by Ore in [25]. These polynomials are of the form $P(z) = \sum_i p_i z^{q^i}$, with $p_i \in \mathbb{F}_{q^m}$ and finitely many nonzero coefficients. We adopt the shorthand notation $[i] := q^i$, so that $P(z) = \sum_i p_i z^{[i]}$.

The set of q-polynomials with coefficients in $\mathbb{F}_{q^m}$, equipped with addition and composition, forms a non-commutative ring denoted by $\mathcal{P}_{q^m}$. These polynomials induce $\mathbb{F}_q$-linear endomorphisms of $\mathbb{F}_{q^m}$.

Definition 7 Let $\mathbf{g} = (g_1, \ldots, g_n) \in \mathbb{F}_{q^m}^n$ be such that the g_j's are $\mathbb{F}_q$-linearly independent, and let $k < n$. The *Gabidulin code* of length n, dimension k, and support $\mathbf{g}$ is defined as:

$$Gab_k(\mathbf{g}) = \left\{(P(g_1), \ldots, P(g_n)) \in \mathbb{F}_{q^m}^n \;\middle|\; P \in \mathcal{P}_{q^m},\ \deg_q(P) < k\right\}.$$

A generator matrix of $Gab_k(\mathbf{g})$ is given by the $k \times n$ matrix:

$$G = \begin{pmatrix} g_1^{[0]} & \cdots & g_n^{[0]} \\ g_1^{[1]} & \cdots & g_n^{[1]} \\ \vdots & \ddots & \vdots \\ g_1^{[k-1]} & \cdots & g_n^{[k-1]} \end{pmatrix}.$$

Gabidulin codes meet the Singleton bound for the rank-metric, and therefore satisfy $d_R = n - k + 1$. Consequently, they can correct up to $\tau_{\max} = \left\lfloor \frac{n-k}{2} \right\rfloor$ rank errors.

Note that the structure of Gabidulin codes is analogous to that of Reed-Solomon codes, with the classical monomial powers g_j^i replaced by Frobenius powers $g_j^{[i]}$, and the support being a set of linearly independent elements over $\mathbb{F}_q$.

It is also well known that a parity-check matrix of a $[n, k, d]$-Gabidulin code is given by:

$$H = \begin{pmatrix} h_1^{[0]} & \cdots & h_n^{[0]} \\ h_1^{[1]} & \cdots & h_n^{[1]} \\ \vdots & \ddots & \vdots \\ h_1^{[d-2]} & \cdots & h_n^{[d-2]} \end{pmatrix},$$

where $\mathbf{h} = (h_1, \ldots, h_n) \in Gab_k(\mathbf{g})^{\perp}$ is a codeword of full rank.

4.2 Rank Generalized Subspace Subcodes of Gabidulin codes

We now specialize the results established previously in the important family of Gabidulin codes. We first present a lemma relating the generalized subspace subcodes of Gabidulin codes to a well-known Gabidulin parent code, and derive explicit cardinality bounds.

Lemma 1 *Let $Gab_k(\mathbf{g})$ be an $[n, k, d]$-Gabidulin code over $\mathbb{F}_{q^m}$. Consider subspaces $V_1, \ldots, V_n$ of dimensions $s_1, \ldots, s_n \leq m$ and define $W = \prod_{i=1}^n V_i$. Then, each codeword of $Gab_k(\mathbf{g}) \cap W$ maps uniquely to a codeword of the Gabidulin code $\mathcal{B}_W$ having parity-check matrix*

$$T = (\beta_j^{[m-i+1]})_{1 \leq i \leq d-1,\, 1 \leq j \leq s}$$

with $s = \max_i(s_i)$ and $(\beta_1, \ldots, \beta_s)$ linearly independent.

Proof According to Proposition 5, we have

$$c \in \mathcal{C} \cap W \Longleftrightarrow \begin{cases} c = (\beta_1, \beta_2, \cdots, \beta_{s_n})\boldsymbol{U} \\ (\beta_1, \beta_2, \cdots, \beta_{s_n})\boldsymbol{U}\boldsymbol{H}^T = 0 \end{cases}$$

where $\boldsymbol{H} = \begin{pmatrix} h_1^{[0]} & \cdots & h_n^{[0]} \\ h_1^{[1]} & \cdots & h_n^{[1]} \\ \vdots & \cdots & \vdots \\ h_1^{[d-2]} & \cdots & h_n^{[d-2]} \end{pmatrix}$.

By setting $\boldsymbol{L} = \boldsymbol{U}\boldsymbol{H}^T$, we have

$$\boldsymbol{L}_{i,j} = \sum_{l=1}^{n} u_{i,l} h_l^{[j-1]} = \left(\sum_{l=1}^{n} u_{i,l} h_l\right)^{[j-1]} = \boldsymbol{L}_{i,1}^{[j-1]}.$$

Since $0 = (\beta_1, \beta_2, \ldots, \beta_{s_n})\boldsymbol{U}\boldsymbol{H}^T$, for $w_i = \boldsymbol{L}_{i,1}$ and $1 \leq j \leq d-1$ we have

$$
\begin{aligned}
0 &= (\beta_1, \beta_2, \ldots, \beta_{s_n}).\begin{pmatrix} w_1^{[j-1]} \\ w_2^{[j-1]} \\ \vdots \\ w_{s_n}^{[j-1]} \end{pmatrix} \\
&= \beta_1 w_1^{[j-1]} + \beta_2 w_2^{[j-1]} + \cdots + \beta_{s_n} w_{s_n}^{[j-1]} \\
&= (\beta_1^{[m-j+1]} w_1 + \beta_2^{[m-j+1]} w_2 + \cdots + \beta_{s_n}^{[m-j+1]} w_{s_n})^{[j-1]} \\
&= \beta_1^{[m-j+1]} w_1 + \beta_2^{[m-j+1]} w_2 + \cdots + \beta_{s_n}^{[m-j+1]} w_{s_n} \\
&= (w_1, \ldots, w_{s_n}) \begin{pmatrix} \beta_1^{[m-j+1]} \\ \beta_2^{[m-j+1]} \\ \vdots \\ \beta_{s_n}^{[m-j+1]} \end{pmatrix}
\end{aligned}
$$

which leads to

$$
(w_1, \ldots, w_{s_n}) \begin{pmatrix} \beta_1^{[m]} & \beta_1^{[m-1]} & \cdots & \beta_1^{[m-d+2]} \\ \beta_2^{[m]} & \beta_2^{[m-1]} & \cdots & \beta_2^{[m-d+2]} \\ \vdots & \vdots & \cdots & \vdots \\ \beta_{s_n}^{[m]} & \beta_{s_n}^{[m-1]} & \cdots & \beta_{s_n}^{[m-d+2]} \end{pmatrix} = 0
$$

Let $\mathcal{B}_W$ be the Gabidulin code having parity-check matrix $\boldsymbol{T} = (\beta_j^{[m-i+1]})_{i=1,j=1}^{d-1,s_n}$. Then, $c = (\beta_1, \beta_2, \ldots, \beta_{s_n})\boldsymbol{U} \in Gab_k(\mathbf{g}) \cap W \Longrightarrow (w_1, \ldots, w_{s_n}) \in \mathcal{B}_W$.

The lemma above allows us to establish explicit cardinality bounds.

Theorem 3 *Let $Gab_k(\mathbf{g})$ be a $[n, k, d]$-Gabidulin code over $\mathbb{F}_{q^m}$, and $W = \prod_i V_i$ as above. Then,*

$$
q^{\sum_i s_i - m(n-k)} \leq |Gab_k(\mathbf{g}) \cap W| \leq q^{m(\max_i(s_i)-(n-k))}.
$$

Proof According to Lemma 1, $c = (\beta_1, \beta_2, \ldots, \beta_{s_n})\boldsymbol{U} \in Gab_k(\mathbf{g}) \cap W \Longrightarrow (w_1, \ldots, w_{s_n}) \in \mathcal{B}_W$. So we have $|Gab_k(\mathbf{g}) \cap W| \leq |\mathcal{B}_W| = q^{m(s_n-d+1)} = q^{m(\max\limits_i(s_i)-d+1)}$. According to Proposition 5, and since $d = n - k + 1$, we get

$$
q^{\sum\limits_{i=1}^{n} s_i - m(n-k)} \leq |Gab_k(\mathbf{g}) \cap W| \leq q^{m(\max\limits_{1 \leq i \leq n}(s_i)-(n-k))}
$$

One can remark that taking $s_1 = \cdots = s_n = s$ gives the following corollary, which is also a known result from [11].

Corollary 4 ([11]) *Let $Gab_k(\mathbf{g})$ be a $[n, k, d]-$Gabidulin code of length n over $\mathbb{F}_{q^m}$ and V be a $\mathbb{F}_q-$subspaces of $\mathbb{F}_{q^m}$ with dimension $s \leq m$. If $ns - m(n-k) > 0$ then,*

$$
q^{ns-m(n-k)} \leq |Gab_k(\mathbf{g})_{|V}| \leq q^{m(s-(n-k))}.
$$

The following result is a direct consequence of Proposition 3, and will be useful to compute a generator matrix of a Generalized Subspace Subcode of a Gabidulin code.

Proposition 6 *The q-ary image $Gab_k(\mathbf{g})(B_i)_i$ of a Gabidulin code $Gab_k(\mathbf{g})$ with respect to the n bases $(B_1, \ldots, B_n)$ of $\mathbb{F}_{q^m}$, is generated by the block matrix $((\boldsymbol{M}_{g_j}^{q^{i-1}}))_{1\leq i\leq k, 1\leq j\leq n} \in (\mathbb{F}_q^{m\times m})^{k\times n}$ where the matrix $\boldsymbol{M}_{g_j}^T = \mathcal{M}_{B,B_j}(\theta_{g_j})$ is the matrix of the linear map $\theta_{g_j} = \phi_{B_j} \circ f_{g_j} \circ \phi_B^{-1}$ with $f_{g_j}(x) = g_j x$ for any $x \in \mathbb{F}_{q^m}$, and B being a fixed basis.*

Using the previous results and the relation between shortening and puncturing operations, we obtain the following explicit algorithm to construct a generator matrix for the generalized subspace subcodes of Gabidulin codes.

Algorithm 1 Generator matrix of $Gab_k(\mathbf{g}) \cap W$

Input A generator matrix $G \in \mathbb{F}_{q^m}^{k\times n}$ of $Gab_k(\mathbf{g})$, a set of n bases $D_1, \ldots, D_n$ of subspaces from $\mathbb{F}_{q^m}$ with dimensions respectively $s_1, \cdots, s_n \leq m$.

Output G_U generator matrix of $(Gab_k(\mathbf{g}) \cap W)(D_j)_j$

$U = \bigcup_{j=1}^{n} u_j$ such that for $1 \leq j \leq n$, $u_j = \{(j-1)m+1, \ldots, (j-1)m+s_j\}$.
Complete the families $D_1, \ldots, D_n$ into n basis $B_1, \ldots, B_n$ of $\mathbb{F}_{q^m}$
Set $\boldsymbol{Q}_j$ be the change-of-basis matrix from the basis B to the basis B_j,
Compute a generator matrix $\hat{\boldsymbol{G}}_B$ of $Gab_k(\mathbf{g})B$
Compute a parity check matrix $\hat{\boldsymbol{H}}_B$ from $\hat{\boldsymbol{G}}_B$
Compute $\hat{\boldsymbol{H}}_{(B_j)_j}^B = \hat{\boldsymbol{H}}_B \cdot \mathrm{Diag}((\boldsymbol{Q}_1^{-1})^T, \ldots, (\boldsymbol{Q}_n^{-1})^T)$
Compute $\boldsymbol{H}_U = (\hat{\boldsymbol{H}}_{(B_j)_j}^B)\bar{U}$ as a parity-check matrix of $(Gab_k(\mathbf{g}))\bar{U}$
Compute G_U, generator matrix of $(Gab_k(\mathbf{g}) \cap W)(D_j)_j$ from $\boldsymbol{H}_U$.

Algorithm 1 thus provides a practical method to explicitly generate generalized subspace subcodes.

5 Parent Codes of Gabidulin RGSS Codes

In the previous section, we established a close relationship between generalized subspace subcodes of Gabidulin codes and certain "parent" Gabidulin codes. We now formalize this relationship by introducing the notion of a *parent code*, which plays a key role for decoding and structural analysis. Recall that given $s = \max(s_i) \geq d$, each codeword in $Gab_k(\mathbf{g}) \cap W$ corresponds uniquely to a codeword in the Gabidulin code $\mathcal{B}_W$, which is an MRD code with parameters $[s, s-d+1, d]$ and parity-check matrix:

$$T = (\beta_j^{[m-i+1]})_{1\leq i\leq d-1,\, 1\leq j\leq s}.$$

Definition 8 (Parent Code) Let $W = \prod_{i=1}^{n} V_i$ be a product of $\mathbb{F}_q$-subspaces of $\mathbb{F}_{q^m}$. The Gabidulin code $\mathcal{B}_W$ defined by the parity-check matrix

$$T = (\beta_j^{[m-i+1]})_{1\leq i\leq d-1,\, 1\leq j\leq s}$$

is called the *parent code* of the generalized subspace subcode $Gab_k(\mathbf{g}) \cap W$, denoted by $P_{Gab_k(\mathbf{g})\cap W}$.

The following proposition formalizes the linear embedding from the generalized subspace subcode into its parent code.

Proposition 7 *Let $Gab_k(\mathbf{g})$ be a $[n, k, d]_{q^m}$-Gabidulin code with a parity-check matrix whose first row is $\mathbf{h} = (h_1, \ldots, h_n)$. Consider subspaces V_i with dimensions $s_i \leq m$, and let $W = \prod_i V_i$. Assume bases $\mathbf{b}_i = \{\beta_1, \ldots, \beta_{s_i}\}$ form an inclusion chain, and let $\mathbf{b} = (\beta_1, \ldots, \beta_{s_n})$. Then the $\mathbb{F}_q$-linear map*

$$f_{\mathbf{b}} : W \to \mathbb{F}_{q^m}^{s_n}, \quad c = \mathbf{b}\boldsymbol{U} \mapsto h\boldsymbol{U}^T,$$

is rank-preserving, injective, and satisfies:

$$f_{\mathbf{b}}(Gab_k(\mathbf{g}) \cap W) \subseteq P_{Gab_k(\mathbf{g})\cap W}.$$

Proof Since the vectors in $\mathbf{h}$ are linearly independent over $\mathbb{F}_q$, $\ker(f_{\mathbf{b}}) = \{0\}$. Therefore, $f_{\mathbf{b}}$ is injective and preserves the rank:

$$w_R(\mathbf{b}\boldsymbol{U}) = rank(\boldsymbol{U}) = w_R(h\boldsymbol{U}^T).$$

By construction, it follows that $f_{\mathbf{b}}(Gab_k(\mathbf{g}) \cap W) \subseteq P_{Gab_k(\mathbf{g})\cap W}$.

Remark 1 From the previous Proposition 7, we can also deduce that the minimum distance of d' of $Gab_k(\mathbf{g}) \cap W$ satisfies $d' \geq d$.

In [10], the authors used this technique to make an algorithm for the encoding and decoding of the subcode on the subspace because there was no simple method to do it. Unfortunately, this method offers the same correction capacity as that of the primary code. So we prefer to use the construction described in the previous section, which allows us to have a direct construction method, even if we still do not have an improvement for the correction capacity.

6 Applications to Cryptography

Cryptography based on rank-metric codes is a very serious alternative to reduce key sizes in code-based cryptography, because the best attacks in rank-metric are exponential with a quadratic exponent, while the best in hamming metric are exponential with a linear exponent. The goal of this section is to show that these codes are potential candidates for rank-metric cryptography in a McEliece setting. Let us recall that the general idea in a McEliece-like cryptosystem is to first choose a linear code $\mathcal{C}$ from a family of structured codes that will serve as the secret key. The code $\mathcal{C}$ will then undergo some transformations in order to hide its structure and will result in a public code $\mathcal{C}_{pub}$ that will be published together with a correction capacity t' depending on the used transformations and the correction capacity of the secret code $\mathcal{C}$. It is well known that, faced with such a system, the attacker must either distinguish the public code from a random code or solve an instance of the general decoding problem with a random code. Following this idea, Gabidulin, Paramonov, and Tretjakov (GPT) [13] were the first to propose the use of the family of Gabidulin codes in a system today known as the GPT cryptosystem. Unfortunately, a polynomial-time attack was proposed on the GPT cryptosystem and its improvements by Overbeck [27] and several other works. The weaknesses mainly come from the fact that Gabidulin codes are invariant under the Frobenius automorphism. To avoid this type of attack, some directions have been taken, and one of them is the use of subcodes of Gabidulin codes. The latter idea was followed by Berger, Gaborit, and Ruatta [1] with a system based on subcodes of $q-$ary images $Gab_k(\mathbf{g})B$ of Gabidulin codes for a fixed basis B. The security of the system is then based on the subcode equivalence problem [1].

According to [1], the complexity for solving the subcode equivalence problem by enumeration of Basis B is a factor of the number of $\mathbb{F}_q-$bases in $\mathbb{F}_{q^m}$. Therefore, Generalized Subspace Subcodes can be used to improve the security or the key sizes of the cryptosystem in [1] as using GSS codes of Gabidulin codes is equivalent to base the system on subcodes of $q-$ary images $Gab_k(\mathbf{g})(B_j)_j$ of Gabidulin codes for a family of bases $B_1, \ldots, B_n$ and, in our case, the attacker must search for n bases instead of just one for solving the subcode equivalence problem by enumeration of Basis. More formally, let's consider $\boldsymbol{Q}, \boldsymbol{Q}_1, \ldots, \boldsymbol{Q}_n \in GL(q, m)$. For a $q-$ary matrix code d[3] we also define $\Phi_{(\boldsymbol{Q}_j)_{j=1}^n, \boldsymbol{Q}}(\mathcal{D})$ by

$$\Phi_{(\boldsymbol{Q}_j)_{j=1}^n, \boldsymbol{Q}}(\mathcal{D}) = \{(\boldsymbol{Q}_1 M_1, \ldots, \boldsymbol{Q}_n M_n)Q \mid M = M_{\phi_B^{-1}}(\mathbf{c}),\ \mathbf{c} \in \mathcal{D}\}. \tag{3}$$

Using Generalized Subspace Subcodes of Gabidulin codes in a McEliece-like cryptosystem results in the following problem.

Problem 1 (Generalized Subcode equivalence on binary Image Codes, GSIC). Given two $q-$ary matrix codes $\mathcal{C}$ and $\mathcal{D}$, find $\boldsymbol{Q}, \boldsymbol{Q}_1, \ldots, \boldsymbol{Q}_n \in GL(q, m)$, such that $\Phi_{(\boldsymbol{Q}_j)_{j=1}^n, \boldsymbol{Q}}(\mathcal{D})$ is a subcode of $\mathcal{C}$.

One can remark that for $\boldsymbol{Q}_1 = \boldsymbol{Q}_2 = \cdots = \boldsymbol{Q}_n$, the previous problem gives rise to the SEMC problem defined in [1, Theorem 3], and shown to be NP-complete.

[3] Here the code d can be viewed as included in $\mathbb{F}_q^{mn}$, but we use the term matrix code here to make connexion with [1] where the codewords of d are viewed as matrices from $\mathbb{F}_q^{m\times n}$, and it is also the case here.

One of the best ways to solve the SEMC as shown in [1], is to enumerate the set of all $\mathbb{F}_q$-bases in $\mathbb{F}_{q^m}$ to find $\boldsymbol{Q}_1$ whereas in our case we need to find n different bases $\boldsymbol{Q}_1, \boldsymbol{Q}_2, \ldots, \boldsymbol{Q}_n$ simultaneously and, these will give a cost estimated as

$$\frac{m^n N_{q,m,n}}{m(q^m-1)} \times \left(n + \frac{q-1}{q}\right) m^3 \times k' m^2 (n-k)$$

instead of

$$\frac{N_{q,m,1}}{m(q^m-1)} \times \left(n + \frac{q-1}{q}\right) m^3 \times k' m^2 (n-k)$$

as obtained in [1], where

$$N_{q,m,n} = \left(\prod_{i=0}^{m-1} (q^m - q^i)\right)^n = N_{q,m,1} \cdot \left(\prod_{i=0}^{m-1} (q^m - q^i)\right)^{n-1}.$$

We emphasize that the idea of this section was not to propose a cryptosystem, because this requires a complete overview of the different existing attacks, and this goes beyond the scope of this paper. However, we wanted to at least show that generalized subspace subcodes in the rank-metric deserve to be studied for their potential applications in cryptography. We see from the above that there could be a significant gain by using RGSS of Gabidulin codes in a McEliece settings compared to the system of Berger et al. [1].

7 Conclusion

In conclusion, we have generalized the notion of Subspace Subcodes in Rank-metric introduced by Gabidulin and Loidreau, and also characterized this family by giving an algorithm allowing to compute generator and parity-check matrices of these codes from the generator and parity-check matrices of the associated extended codes. We have also studied the specific case of generalized subspace subcodes of Gabidulin codes and show that they are applicable to cryptography in a McEliece setting.

The proposed algorithm for generating generator and parity-check matrices based on associated extended codes not only enhances the practicality of these codes but also opens avenues for further exploration and application in cryptology.

References

1. Berger, T.P., Gaborit, P., Ruatta, O.: Gabidulin matrix codes and their application to small ciphertext size cryptosystems. In: Patra, A., Smart, N.P. (eds.) Progress in Cryptology—INDOCRYPT 2017, pp. 247–266. Springer International Publishing, Cham (2017)

2. Berger, T.P., Gueye, C.T., Klamti, J.B., Ruatta, O.: Designing a public key cryptosystem based on quasi-cyclic subspace subcodes of Reed-Solomon codes. In: Gueye, C.T., Persichetti, E., Cayrel, P.-L., Buchmann, J. (eds.) Algebra, Codes and Cryptology, pp. 97–113. Springer International Publishing, Cham (2019)
3. Berger, T.P., Gueye, C.T., Klamti, J.B.: Generalized subspace subcodes with application in cryptology. IEEE Trans. Inf. Theory **65**(8), 4641–4657 (2019)
4. Berger, T.P., Loidreau, P.: How to mask the structure of codes for a cryptographic use. Des. Codes Crypt. **35**, 63–79 (2005)
5. Couvreur, A., Lequesne, M.: On the security of subspace subcodes of Reed-Solomon codes for public key encryption. IEEE Trans. Inf. Theory **68**(1), 632–648 (2021). Oct.
6. Couvreur, A., Otmani, A., Tillich, J.-P.: Polynomial time attack on wild McEliece over quadratic extensions. IEEE Trans. Inf. Theory **63**(1), 404–427 (2017)
7. Delsarte, P.: Bilinear forms over a finite field, with applications to coding theory. J. Combin. Theory Ser. A **25**(3), 226–241 (1978)
8. El Qachchach, I., Habachi, O., Cances, J.-P., Meghdadi, V.: Efficient multi-source network coding using low rank parity check code. In: 2018 IEEE Wireless Communications and Networking Conference (WCNC), pp. 1–6. IEEE (2018)
9. Gabidulin, E.M.: Theory of codes with maximum rank distance. Problemy Peredachi Informatsii **21**(1), 3–16 (1985)
10. Gabidulin, E.M., Loidreau, P.: On subcodes of codes in rank metric. In: Proceedings of the International Symposium on Information Theory (ISIT 2005), pp. 121–123 (2005)
11. Gabidulin, E.M., Loidreau, P.: Properties of subspace subcodes of Gabidulin codes. Adv. Math. Commun. **2**(2), 147–157 (2008)
12. Gabidulin, E.M.: Attacks and counter-attacks on the GPT public key cryptosystem. Des. Codes Crypt. **48**(2), 171–177 (2008)
13. Gabidulin, E.M., Paramonov, A.V., Tretjakov, O.V.: Ideals over a non-commutative ring and their application in cryptology. In: Workshop on the Theory and Application of Cryptographic Techniques, pp. 482–489. Springer (1991)
14. Gibson, K.: Severely denting the Gabidulin version of the McEliece public key cryptosystem. Des. Codes Crypt. **6**(1), 37–45 (1995)
15. Gibson, K.: The security of the Gabidulin public key cryptosystem. In: Maurer, U. (ed.) Advances in Cryptology—EUROCRYPT '96, vol. 1070 of Lecture Notes in Computer Science, pp. 212–223. Springer (1996)
16. Hattori, M., McEliece, R.J., Solomon, G.: Subspace subcodes of Reed-Solomon codes. IEEE Trans. Inf. Theory **44**(5), 1861–1880 (1998)
17. Horlemann-Trautmann, A.-L., Marshall, K., Rosenthal, J.: Extension of Overbeck's attack for Gabidulin-based cryptosystems. Des. Codes Crypt. **86**(2), 319–340 (2018)
18. Huffman, W.C., Pless, V.: Fundamentals of Error-Correcting Codes. Cambridge University Press (2003)
19. Jensen, J.M.: Subgroup subcodes. IEEE Trans. Inf. Theory **41**(3), 781–785 (1995)
20. Kalachi, H.T.: On the failure of the smart approach of the GPT cryptosystem. Cryptologia **46**(2), 167–182 (2022)
21. Loidreau, P.: Designing a rank metric based McEliece cryptosystem. In: Proceedings of the International Workshop on Post-Quantum Cryptography, pp. 142–152. Springer (2010)
22. Lu, H.-F., Kumar, P.V.: A unified construction of space-time codes with optimal rate-diversity tradeoff. IEEE Trans. Inf. Theory **51**(5), 1709–1730 (2005)
23. Melchor, C.A., Aragon, N., Bardet, M., Bettaieb, S., Bidoux, L., Blazy, O., Deneuville, J.-C.: ROLLO-Rank-Ouroboros, LAKE & LOCKER. NIST Competition for Post-Quantum Cryptography (2019)
24. Melchor, C.A., Aragon, N., Bettaieb, S., Bidoux, L., Blazy, O., Deneuville, J.-C., Gaborit, P., Zémor, G.: Rank Quasi-Cyclic (RQC). NIST Competition for Post-Quantum Cryptography (2017)
25. Ore, Ø.: On a special class of polynomials. Trans. Am. Math. Soc. **35**(3), 559–584 (1933)

26. Otmani, A., Kalachi, H.T., Ndjeya, S.: Improved cryptanalysis of rank metric schemes based on Gabidulin codes. Des. Codes Crypt. **86**(9), 1983–1996 (2018)
27. Overbeck, R.: A new structural attack for GPT and variants. In: Proceedings of the International Conference on Cryptology in Malaysia, pp. 50–63. Springer (2005)
28. Rashwan, H., Gabidulin, E.M., Honary, B.: A smart approach for GPT cryptosystem based on rank codes. In: 2010 IEEE International Symposium on Information Theory, pp. 2463–2467. IEEE (2010)
29. Rashwan, H., Gabidulin, E.M., Honary, B.: Security of the GPT cryptosystem and its applications to cryptography. Secur. Commun. Netw. **4**(8), 937–946 (2011)
30. Reed, I.S., Solomon, G.: Polynomial codes over certain finite fields. J. Soc. Ind. Appl. Math. **8**, 300–304 (1960)
31. Shor, P.W.: Algorithms for quantum computation: discrete logarithms and factoring. In: Goldwasser, S. (ed.) Proceedings of the 35th Annual Symposium on Foundations of Computer Science (FOCS), pp. 124–134 (1994)
32. Wu, Y.: On expanded cyclic and Reed-Solomon codes. IEEE Trans. Inf. Theory **57**(2), 601–620 (2011)

Proxy Cryptography with Post-Quantum Probabilistic Full Domain Hash

Bacar Nourdine, Mouhamed Lamine Mbaye, and Demba Sow

Abstract Proxy cryptography is a valuable tool in signature schemes, enabling a user to delegate signing capabilities to another entity in a controlled manner. Proxy functions can transform a standard signature scheme into a unidirectional variant, where the security of the resulting construction is directly linked to that of the base scheme. Our scheme introduces a random salt for each signing process and incorporates it into the computation of the encryption exponent. The construction achieves a tighter security reduction in the random oracle model, under the assumption that inverting RSA remains hard. While the scheme inherits the limitations of RSA with respect to Shor's algorithm, it strengthens practical security in pre-quantum and transitional environments. We show how the unidirectional proxy function can be applied to the enhanced pqPFDH to obtain a generic unidirectional signature scheme that is existentially unforgeable under chosen message attacks (UF-CMA).

Keywords Signature scheme · Proxy function · Full domain hash · Post-quantum RSA · Security reduction · Chosen message attack

1 Introduction

One of the fundamental roles that cryptography plays in the security of data transmission lies in the digital signature process which guarantees authentication, data integrity, and non-repudiation. Since the discovery of the RSA signing system [25], research in this area has steadily increased and has generally led to the establishment of signing techniques with significant advantages in terms of functionality. In this context, Bellare and Rogaway have developed a process for constructing signature schemes whose security is relative to that of the RSA. In 1993, they introduced

B. Nourdine (✉) · M. L. Mbaye · D. Sow
Université Cheikh Anta Diop de Dakar, Faculté des Sciences et Techniques, Dakar, Senegal
e-mail: nourdinbacar@gmail.com

D. Sow
e-mail: demba1.sow@ucad.edu.sn

© The Author(s), under exclusive license to Springer Nature Switzerland AG 2026
C. T. Gueye et al. (eds.), *Mathematics, Computer Science, Cybersecurity and Artificial Intelligence*, Springer Proceedings in Mathematics & Statistics 536,
https://doi.org/10.1007/978-3-032-17666-0_6

the Full Domain Hash signature scheme and the Probabilistic Signature Scheme [5] which were proven secure against chosen message attack in the random oracle model [4] assuming that inverting RSA is hard. The gap in this eminent work is that the security tightness of the FDH signature scheme can be enhanced. Then, it was revisited by other cryptographers like Coron [10, 11], in order to strengthen the tightness security by exhibiting different proof which provides tighter security reduction. Thus, research efforts directed toward the establishment of secure signature schemes remain more essential than ever and further contribute to confirming the preponderant role that cryptography plays in the process of securing data and information.

Another method of producing secure signatures is to use the notion of atomic proxy cryptography introduced in 1998 by Matt Blazz and Martin Strauss [1, 2]. They highlighted the notion of proxy functions which are tools to convert a valid signature for one key into another valid for another key without disclosing the secret signature keys. This signing technical was first explained by Mambo et al. [22], and allows for delegating signature rights which is an idea that is the subject of several types of research [6, 19–21]. It's true that the objective of these seminal works can differ in terms of primitive construction but the base always remains threshold cryptography which is still a very well updated research field. This is illustrated by the fact that a project to drive an effort to standardize threshold schemes for cryptographic primitives has been established by the Computer Security Division at the National Institute of Standards and Technology (NIST) [26]. But, in this work, we use the notion of unidirectional proxy function [16] which is a category of proxy functions that allow one user to generate signature corresponding to the secret key of another user even if the first user doesn't hold the secret key.

Another RSA-based signature scheme called post-quantum Probabilistic Full Domain Hash (pqPFDH) [23] is a main idea in this work. It's a variant of the Full domain Hash signature scheme with a random generated for each signature process. Its security, relatively close to that of RSA, was proven in the random oracle model. In the security proof, it is assumed that the hash function used is ideal and the RSA trapdoor permutation holds.

The goal, in this paper, is to construct the unidirectional version of pqPFDH which is proven secure in the random oracle model. The security is not only guaranteed against any user U but also against the proxy P and the user F that hold two secret parts delegated by U from its private key so that they can securely sign on behalf of him. This feature allows F and P to cooperate and generate signatures corresponding to the secret key of U but they can't perform this functionality without collaboration. This construction can be very important in the functioning of administrations. Indeed, an agent may be called upon to sign important transactions without being bound by the required secrecy. It's a good idea that the owner of the secret can delegate his signing rights by giving a proxy key to this representative without the risk of compromising security.

To prove the security of signature scheme against the proxy P, the users F, and U, we generally proceeds by demonstrating that, if a polynomial-time adversary $\mathcal{A}$ can break the signature scheme, it can be used by a reduction algorithm $\mathcal{R}$ to invert in polynomial time some related one-way-function. Given an attacker $\mathcal{A}$ which can break the signature in time $\tau_{\mathcal{A}}$ with success probability at least $\varepsilon_{\mathcal{A}}$ for the reduction proof, $\mathcal{R}$ must simulate the environment of $\mathcal{A}$ and solve the problem (invert the one way function) with time $\tau_{\mathcal{R}} \geq \tau_{\mathcal{A}}$ and success probability $\varepsilon_{\mathcal{R}} \leq \varepsilon_{\mathcal{A}}$. For tightness of the reduction, it is required to have $\varepsilon_{\mathcal{R}} \approx \varepsilon_{\mathcal{A}}$ and $\tau_{\mathcal{R}} \approx \tau_{\mathcal{A}} + polynom(k)$, where k is a security parameter.

The paper is structured as follows. The next section describes some previous works on signature scheme based on RSA and their security. It also includes some terms on the notion of threshold cryptography and particularly the concept of proxy cryptography. In Sect. 3, we present some preliminaries which help to understand the rest of the paper (security model in 3.1, signature scheme in 3.2, unidirectional signature proxy function in 3.3, unidirectional generic signature scheme in 3.4 and post-quantum RSA problem 3.5). Section 4 prescribes the pqPFDH scheme which is a post-quantum version of the PFDH signature scheme with a generated random allowing to compute the RSA exponent for each signature process. In Sect. 5, we produce the unidirectional version of the pqPFDH signature scheme and a proof of security on this construction.

2 Related Works

In order to strengthen the basic RSA signature, Bellare and Rogaway proposed in [5] the Full Domain Hash signature scheme. Assuming that the used hash function is ideal, they proved that their scheme is secure against chosen message attack in the random oracle model [4]. If f is a trapdoor permutation and $\mathcal{RO}$ is a random function from $\{0, 1\}^*$ to the domain of f, they proved that signing a message m via $f^{-1}(\mathcal{RO}(m))$ is secure and proposed a security reduction for RSA-FDH where the reduction algorithm $\mathcal{R}$ will provide a perfect simulation and $(\varepsilon_{\mathcal{R}}, \tau_{\mathcal{R}})$-solves RSA trapdoor permutation with success probability $\varepsilon_{\mathcal{R}} \geq \frac{\varepsilon_{\mathcal{A}}}{q_h+q_{sig}+1}$ and time bound $\tau_{\mathcal{R}} \leq \tau_{\mathcal{A}} + (q_h + q_{sig} + 1)polynom(k)$, where q_h and q_{sig}) are respectively the number of hash queries and signature queries made by the attacker.

To improve the tightness of RSA-FDH, in 2002, Coron [10], proposed a security reduction where the reduction algorithm $\mathcal{R}$ will provide a perfect simulation and $(\varepsilon_{\mathcal{R}}, \tau_{\mathcal{R}})$-solves RSA trapdoor permutation with success probability $\varepsilon_{\mathcal{R}} = \frac{\varepsilon_{\mathcal{A}}}{exp(1)q_{sig}}$ and time bound $\tau_{\mathcal{R}} = \tau_{\mathcal{A}} + (q_h + q_{sig} + 1)\mathcal{O}(k^3)$, where q_h (respectively q_{sig}) is the number of queries made by the attacker to the oracle hash (respectively to the oracle signature).

Coron studied also the tightness of probabilistic FDH (PFDH briefly) in [11], and proposed a tight security reduction where the reduction algorithm $\mathcal{R}$ will provide a perfect simulation and $(\varepsilon_{\mathcal{R}}, \tau_{\mathcal{R}})$-solves RSA trapdoor permutation with success probability $\varepsilon_{\mathcal{R}} = \dfrac{\varepsilon_{\mathcal{A}}}{(1 + 6\frac{q_{sig}}{2^{k_0}})}$ and time bound $\tau_{\mathcal{R}} = \tau_{\mathcal{A}} + (q_h + q_{sig})\mathcal{O}(k^3)$ where

q_h (respectively q_{sig}) is the number of queries made by the attacker to the oracle hash (respectively to the oracle signature).

In 2001, in the case of proxy cryptography, Bellare and Sandhu proposed in [6] the two-party RSA signature scheme which allows a client and a server, each holding a shared secret, to collaborate and compute an RSA signature under the corresponding public key. In this same year, H. Kim et al. applied in [19] proxy signature scheme to the mobile agent system to enhance security and efficiency. Here, a customer generates delegation key pair and allows the server to sign with it on behalf of him.

In 2002, Dodis and Reyzin [12], generalizing Coron's work, showed that a similar result holds for any trapdoor permutation induced by a family of claw-free permutations. They proved that a tight security reduction is impossible for RSA-FDH, RSA-PFDH, and PSS-R with small random. Moreover, they showed that, this cases, for any signature scheme outputs $sign(m) = \left(f^{-1}(\mathcal{RO}(m))\right)$ or $sign_r(m) = \left(f^{-1}(\mathcal{RO}(m, r)), r\right)$, where f^{-1} is the inverse of the trapdoor permutation, m is the message and r is a random, if the scheme is to be analyzed with a general "black-box" trapdoor permutation f.

In 2005, Dodis et al. [13], using previous work of Dodis and al., proved that one can't hope to prove $sign(m) = \left(f^{-1}(\mathcal{RO}(m))\right)$ secure under any assumption which is satisfied by random permutations. Pietrzak tells that their work does not mean that RSA-FDH with SHA1 is insecure, but it is just impossible to prove it (with a tight security reduction).

In 2009, Bégueline and Grégoire [3] used a general framework called CertiCrypt, allowing to formalize exact security proofs of cryptographic systems in the computational model, to prove the existential unforgeability under adaptive chosen-message attacks of the FDH and PFDH.

In 2011, Boneh et al. [7] proved that FDH and PFDH signature schemes with preimage sampleable Trapdoor Functions have a history-free reduction. This means that their security in the classical random oracle model implies their security the quantum accessible random oracle model. They also proved the quantum security of a variant of FDH due to Katz and Wang [17] which has tight security reduction.

In 2018, Kiltz and Kakvi [18] revisited Coron's work and contradict the fact that Coron said that the security loss of q_S in the proof of FDH is optimal and cannot possibly be improved. They showed that it only holds if the underlying trapdoor permutation is certified. They also use a stronger assumption called Phi-Hiding assumption introduced by Cachin et al. [8] to give a new tight security reduction of the FDH signature scheme.

In 2020, Yan Hue et al. proposed in [15] a signature scheme called the Easy Simple Factoring-based (ESF) signature scheme proven secure in the random oracle model. Compared to the FDH, their scheme has a more powerful efficiency performance in signing and verifying algorithms in the case that public keys are fixed to small bit lengths.

In 2022, M. L. Mbaye et al. proposed in [23] a signature scheme called pqPFDH based on the post quantum RSA principle. The idea here is to make algorithms that we can use to sign and verify messages in a quantum area.

3 Preliminaries

In this section, we recall some definitions and known results about signatures. Definitions, basic notations, and classical results are followed in "Introduction to Modern Cryptography" [17] of Katz and Lindell, and in "Provable Security for Public Key Schemes" of Pointcheval [24].

3.1 Security Model

Random Oracle Model: For any constant k, a random oracle is a function F_{rand} selected randomly in the set $\mathcal{F}_k$ of functions from $\{0, 1\}^*$ to $\{0, 1\}^k$.

Proof in the Random Oracle Model: Proof in the Random Oracle Model is described in [4, 9].

- Suppose that the hash function is a random function, i.e., in the simulation process, the hash function is replaced by a random oracle which outputs a random value for each new input.
- The only way to compute the hash function is to query the oracle hash.
- The reduction algorithm $\mathcal{R}$ must simulate the environment of the attacker $\mathcal{A}$ with her public key only.
- When the attacker $\mathcal{A}$ requests the oracle hash, the reduction algorithm $\mathcal{R}$ can choose the random to return as digest; hence $\mathcal{R}$ is able to embed the challenge (any information which able $\mathcal{R}$ to invert the related one way function at the end of the game) of his choice in the answer of a hash oracle query to $\mathcal{A}$.
- At the end of the simulation, if the attacker $\mathcal{A}$ outputs a valid forgery (which be never returned by the oracle signature) then $\mathcal{R}$ must be able to invert the related one-way function with good tightness.

3.2 Signature Scheme

A Randomized signature scheme is a tuple of three algorithms $\mathcal{S} = (\mathcal{KG}, \mathcal{SIG}, \mathcal{VER})$ described by as follows:

- Key Generation algorithm ($\mathcal{KG}$): with input a security parameter k, the $\mathcal{KG}$ algorithm outputs a pair of keys (pub_{key}, sec_{key}).
- Signature Algorithm ($\mathcal{SIG}$):
  - with input a security parameter k, the signing algorithm produces a random r.
  - with input (sec_{key}, m, r), the $\mathcal{SIG}$ outputs a signature σ.
- Verification Algorithm ($\mathcal{VER}$): with input (m, σ, pub_{key}), the $\mathcal{VER}$ algorithm returns 1 if the signature is valid, and 0 otherwise.

Security of Randomized Signature Schemes: Goldwasser, Micali and Rivest, in [14], introduced the basic security notion for signatures called "existential unforgeability with respect to adaptive chosen-message attacks".

For this, a reduction algorithm $\mathcal{R}$ and an attacker $\mathcal{A}$ simulate the following game.

Setup: $\mathcal{R}$ runs the algorithm $\mathcal{SIG}$ with a security parameter k as input, to obtain the public key pub_{key} and the secret key sec_{key}, and gives pub_{key} to the adversary.

Queries: Proceeding adaptively, $\mathcal{A}$ may request a signature on any message $m \in \mathcal{M}$ (multiple requests of the same message are allowed) and $\mathcal{R}$ will respond with (m, r, σ) where $\sigma = \mathcal{SIG}(sec_{key}, m, r)$ and r is a random. Let $Hist(\mathcal{S})$ be the signing data base (=set of signatures already outputted by the oracle signature to the queries of the $\mathcal{A}$).

Output: Eventually, $\mathcal{A}$ will output a pair (m, r, σ) and is said to win the game if $\mathcal{VER}(pub_{key}, m, r, \sigma) = 1$ and if $(m, r, \sigma) \notin Hist(\mathcal{S})$ (this last condition forces the attacker $\mathcal{A}$ to output his own forgery).

The probability that $\mathcal{A}$ wins in the above game is denoted $Adv[\mathcal{A}, \mathcal{S}]$.

Unforgeability Against Adaptive Chosen Message Attacks (EUF-CMA): A signature scheme $\mathcal{S} = (\mathcal{KG}, \mathcal{SIG}, \mathcal{VER})$ is existentially unforgeable with respect to adaptive chosen message attacks if for all probabilistic polynomial time attacker $\mathcal{A}$, $Adv[\mathcal{A}, \mathcal{S}]$ is negligible in the security parameter k.

BR-CMA: This adversary model is CMA where the number of random used by the probabilistic signature algorithm is fixed, says D. Hence the signer cannot sign (and outputs distinct values) the same message more than D times.

3.3 Unidirectional Signature Proxy Function

An unidirectional signature proxy function is defined as a tuple $\mathcal{S} = (UniGen, UniSig, UniVer, PSig, FSig)$. For every general user U, the generation algorithm $UniGen(1^k)$ outputs a tuple of keys (sk, vk), where sk is the signing key and vk is the verifying key. Then, sk is used to generate the two proxy keys sk_p and sk_f for the cooperative parties P and F respectively. The $UniSig$ algorithm allows to sign a message m using the key sk ($\sigma = UniSig(m, sk)$). The signature is formed by the couple (m, σ). The verification algorithm $UniVer$ uses the public key vk to verify if a given signature (m, σ) is valid. It outputs 1 (for success) if the signature is valid and 0 otherwise. The correctness property requires that $(UniVer(UniSig(m, sk), vk) = 1)$. This kind of signing is known as the technical standard to sign a message m. Whenever two cooperative parties P and F must generate a signature on behalf of a user U, they have each one their own secret. F uses the function $FSig$ to generate the partial signature of m based on sk_f and asks the proxy P for help. P uses the other function $PSig$ to complete this partial signature into a valid signature.

In terms of security, neither entity (proxy P, user F, user U) can generate alone valid signatures under key sk, even if they know sk_p or sk_f and any of the available public information. The definition below gives more illustration.

Definition 1 Let $\mathcal{S} = (UniGen, UniSig, UniVer, PSig, FSig)$ be a unidirectional signature scheme.
– $\mathcal{S}$ is UF-CMA against the proxy P if $|Succ_{P,\mathcal{S}}(1^k)|$ is negligible, the proxy P is not allowed to ask the $FSig$ oracle for $UniSig(m)$ and $|Succ_{P,\mathcal{S}}$ is defined as below
$|Succ_{P,\mathcal{S}} = Pr[UniVer(m, s) = 1|(sk, vk) \leftarrow UniGen(1^k), (m, s) \leftarrow P^{FSig}(sk_p, vk)]$
– $\mathcal{S}$ is UF-CMA against the user F if $|Succ_{F,\mathcal{S}}(1^k)|$ is negligible, the user F is not allowed to ask the signature oracle $FSig$ oracle for $UniSig(m)$ and $|Succ_{F,\mathcal{S}}$ is defined as below
$|Succ_{F,\mathcal{S}} = Pr[UniVer(m, s) = 1|(sk, vk) \leftarrow UniGen(1^k), (m, s) \leftarrow F^{UniSig}(sk_f, vk)]$
– $\mathcal{S}$ is UF-CMA against the user U if $|Succ_{U,\mathcal{S}}(1^k)|$ is negligible for any PPT adversary U, the user U is not allowed to ask the signature oracle for $UniSig(m)$ and $|Succ_{U,\mathcal{S}}$ is defined as below
$|Succ_{U,\mathcal{S}} = Pr[UniVer(m, s) = 1|(sk, vk) \leftarrow UniGen(1^k), (m, s) \leftarrow U^{UiSig}(vk)]$.

3.4 Unidirectional Generic Signature Scheme

Let $\mathcal{S} = (Gen, Sig, Ver)$ be a standard signature scheme. The corresponding unidirectional generic signature scheme is defined as $\mathcal{S}' = (UniGen, UniSig, UniVer, PSig, FSig)$. The generation algorithm $UniGen$ generates two pairs of keys $((sk_1, vk_1), (sk_2, vk_2))$ for each user U. The secret key for the unidirectional signature scheme is $sk = (sk_1, sk_2)$. The proxy P gets (vk_1, vk_2, sk_1) and the user F gets (vk_1, vk_2, sk_2). The signature algorithm $UniSig$ generates a valid signature for a message m by applying the signature Sig twice: $s = (s_1, s_2) = Sig_1(m)Sig_2(m)$. Similarly, the verification algorithm $UniVer$ verifies whether the signatures generated by $UniSig$ are valid by applying the original verification algorithm Ver algorithm twice: $Ver_1(s_1)Ver_2(s_2)$. The proxy P uses $PSig$ to generate part of the total unidirectional signature: $Sig_1(m)$, while the user F uses $FSig = Sig_2(m)$ to generate the entire signature together with $PSig$.

The following theorem ensures the security of a unidirectional signature scheme relative to the standard scheme to which it is delivered (see [16]).

Theorem 1 *Let $\mathcal{S} = (Gen, Sig, Ver)$ be a standard signature scheme and $\mathcal{S}' = (UniGen, UniSig, UniVer, PSig, FSig)$ the corresponding unidirectional one. If $\mathcal{S}$ is UF-CMA, then $\mathcal{S}'$ is UF-CMA against the proxy P, the user F and the user U.*

3.5 Post Quantum RSA

Post-quantum RSA consists of adjusting RSA parameters to make extremely large keys that will resist the power of quantum computers. The idea is to use many small primes which constitute the secret key and their product gives the public key. Clearly, a user generates K primes $q_1, q_2, ..., q_K$ of the form $q_i = 2p_i^{\beta_i} + 1$, where $1 \leq i \leq K$ and establishes $n = \prod_{i=0}^{K} q_i$.

Now, for a given pqRSA modulus n, an integer e coprime with $\varphi(n) = (q_1 - 1)(q_2 - 1) \cdots (q_K - 1)$ and $z \in (\frac{\mathbb{Z}}{n\mathbb{Z}})^*$, find x such that $x^e = z \mod n$, is the pqRSA problem. Then, an algorithm $\mathcal{R}$ is said to $(\tau_{\mathcal{R}}, \varepsilon_{\mathcal{R}})$-solve the pqRSA problem, if in at most $\tau_{\mathcal{R}}$ operations, $Pr\{(n, e) \leftarrow RSA(1^k), z \leftarrow (\frac{\mathbb{Z}}{n\mathbb{Z}})^*, x \leftarrow \mathcal{R}(n, e, z), x^e = z \mod n\} \geq \varepsilon_{\mathcal{R}}$, where the probability is taken over the distribution of (n, z) and over $\mathcal{R}$'s random tapes.

4 The Post-Quantum Probabilistic Full Domain Hash Signature Scheme

The pqPFDH signature scheme is EUF-BR-CMA secure under the RSA assumption in the random oracle model. It is similar to the Probabilistic Full Domain Hash, but with a random exponent generated for each signature process and used to compute the encryption exponent.

4.1 Signature Process

Here, we elucidate the different steps to follow to generate a signature according to our scheme.

Key Generation Algorithm $\mathcal{KG}$: The key generation algorithm works as follows:

1. With input 1^k, $\mathcal{KG}$ generates a list of K primes $q_1, q_2, \ldots, q_K$ of the form $q_i = 2p_i^{\beta_i} + 1$ for $1 \leq i \leq K$ and computes the modulus $n = q_1 \times q_2 \times \cdots \times q_K$. Then the public key is $p_{key} = n$ and the secret key is $s_{key} = (q_1, q_2, \ldots, q_K)$.
2. It also uses a function G which, on input a random $a \in \{0, 1\}^{k_a}$, outputs an odd integer e which is used to compute d such that $e \cdot d = 1 \ mod \ \varphi(n)$. $G : \{0, 1\}^{k_a} \longmapsto \{0, 1\}^{k_e} \cap \mathbb{N}_{odd}$, where $\{0, 1\}^{k_e}$ is the set of bits representation of the integer e such that $e.d = 1 \ mod \ \varphi(n)$.
3. After concatenated the message m with the random a and the exponent e, one uses the hash function $H : \{0, 1\}^{k_a+k_m+k_e} \longmapsto \mathbb{Z}_n^*$, where k_m is the length of the message m.

Remark 1 To build G, one can proceed as follows: Let $G' : \{0, 1\}^{k_a} \rightarrow \{0, 1\}^{k_e-2}$ be a hash function where k_e is a security parameter, define the hash function G by $G(x) = 1||G'(x)||1 \in [0, n-1] \cap [2^{k_e}, 2^{k_e+1}[\cap \mathbb{N}_{odd}$, where $\mathbb{N}_{odd}$ is the set off odd integers.

Remark 2 Since $G(x)$ is odd then, with a high probability, $G(x)$ is co-prime with $\varphi(n) = 2^K p_1^{\beta_1} p_2^{\beta_2} \ldots p_K^{\beta_K}$ because finding an odd integer which is not co-prime with $\varphi(n)$ is equivalent to factor n (which is known to be difficult).

Signature Algorithm $\mathcal{SIG}(m, s_{key})$*:* To sign a message m, the signer Bob should do the following steps:

(1) pick the private key $(q_1, q_2, \ldots, q_K)$ and the two hash functions G and H;
(2) select a random salt $a \in \{0, 1\}^{k_a}$, compute $e = G(a)$ and $d = e^{-1} \mod \varphi(n)$;
(3) compute $y \longleftarrow H(e||m||a)$ and $\sigma \longleftarrow y^d \mod n$;
(4) The signature of m is (m, a, σ).

Verification algorithm $\mathcal{VER}(p_{key}, (m, a, \sigma))$*:* To verify the signature (m, a, σ) submitted by Bob, the verifier Alice should do the following steps:

(1) pick n, Bob's public key and the signature (m, a, σ);
(2) compute $e = G(a)$ and $H(e||m||a) = y$;
(3) if $y = \sigma^e \mod n$, then return 1; else return 0.

4.2 *Security Proof: Reduction in the Random Oracle Model*

Assuming that inverting RSA is a hard problem, the following theorem proves the security of the pqPFDH signature in the random oracle model. It shows that with pqPFDH, we obtain tight security proof using much shorter random (a random of 5 bits).

Theorem 2 *If there exists an attacker* $\mathcal{A}$ *that* $(q_H, q_G, q_{sig}, \tau_\mathcal{A}, \varepsilon_\mathcal{A})$*-solves EUF-BR-CMA the pqPFDH signature scheme, then there exists a reduction* $\mathcal{R}$ *simulating the environment of* $\mathcal{A}$ *in the random oracle model that* $(\tau_\mathcal{R}, \varepsilon_\mathcal{R})$*-solves RSA with success probability* $\varepsilon_\mathcal{R} = \varepsilon_\mathcal{A}(1 - \frac{1}{2^{|a|}})$ *and time bound* $\tau_\mathcal{R} \leq \tau_\mathcal{A} + q_G\mathcal{O}(1) + (q_H + q_{sig} + 2)\mathcal{O}(k^3)$*, where* $|a|$ *is the size of random used for exponentiation,* $\mathcal{R}$ *receives* q_{sig} *signature queries,* q_H *and* q_G *queries respectively for the hash oracles H and G from* $\mathcal{A}$ *and k is a security parameter.*

5 Unidirectional pqPFDH Signature Scheme

In this section, we show how one user splits its private key in two parts and shares them in order to allow a designated corresponding to sign a message on behalf of him.

Let $\mathcal{S} = (\mathcal{GEN}, \mathcal{SIG}, \mathcal{VER})$ be the pqPFDH signature scheme. The $\mathcal{GEN}$ algorithm generates a random $r \in \{0,1\}^{k_r}$ and uses the function $G : \{0,1\}^{k_a} \longmapsto \{0,1\}^{k_e} \cap \mathbb{N}_{odd}$ to compute e and d. The public key is the couple (n, e) and the secret key is $(d, \phi(n))$. The signature σ of a message m is obtained by performing the operation $\sigma = \mathcal{SIG}(m, d) = H(r||e||m)^d \mod n$, where $H : \{0,1\}^{k_r+k_m+k_e} \longmapsto \mathbb{Z}_n^*$ is an ideal hash function. For a given valid signature (m, r, σ), the verification algorithm $\mathcal{VER}((m, r, \sigma), e))$ returns 1 and 0 if this given signature is not valid.

Then this pqPFDH signature scheme can be transformed into a unidirectional signature scheme $\mathcal{S}' = (\mathcal{UGEN}, \mathcal{USIG}, \mathcal{UVER}, \mathcal{PSIG}, \mathcal{FSIG})$ by the following steps. $\mathcal{UGEN}$ is the key generation algorithm. It uses $\mathcal{GEN}$ to generate the private key d and splits it into parts d_1 and d_2 such $d = d_1 + d_2 \mod \phi(n)$. Then, $d_1 = sk_P$ becomes the secret key of the proxy P and $d_2 = sk_F$ the secret key of the user F. The user U, on his side, must be able to generate a valid signature in a secure way, F and P must cooperate to be able to generate a valid and secure signature. Concerning U, the signature procedure is the same as at the level of the standard schema. It uses $\mathcal{USIG}$ algorithms and $\mathcal{UVER}$ which are identical to the original ones $\mathcal{SIG}$ and $\mathcal{VER}$. The cooperation will take place in case U assigns its signing rights to P and F. In this case, if F wants to sign a message on behalf of U, it uses the $\mathcal{FSIG}$ function to generate one part the unidirectional pqPFDH signature by computing $\sigma_1 = \mathcal{FSIG}(m, sk_F) = \mathcal{SIG}(m, d_2)$. Then, it transfers the pair (m, σ_1) to P which, in turn, completes the signature by performing $\sigma_2 = \mathcal{PSIG}(m, sk_P) = \mathcal{SIG}(m, d_1)$. So, the product $\sigma_1 \cdot \sigma_2$ is the required result since $\sigma = \sigma_1 \cdot \sigma_2 = y^{d_1} \cdot y^{d_2} = y^{d_1+d_2} = y^d$, where $y = H(r||e||m)$.

The following theorem proves the security of the unidirectional pqPFDH signature scheme based on that of the pqPFDH.

Theorem 3 *Let $\mathcal{S} = (\mathcal{GEN}, \mathcal{SIG}, \mathcal{VER})$ be the pqPFDH signature scheme and $\mathcal{S}' = (\mathcal{UGEN}, \mathcal{USIG}, \mathcal{UVER}, \mathcal{PSIG}, \mathcal{FSIG})$ its unidirectional version. Then, $\mathcal{S}'$ is UF-CMA against the proxy P, the user F, and the user U.*

According to Theorem 1, this constructed unidirectional signature scheme is secure against the proxy P, the user F, and the user U.

Proof 1. Consider that the scheme $\mathcal{S}$ is not secure against the proxy P. This means that the quantity $|Success_{P,\mathcal{S}}|$ is negligible. Then, based on P, we can construct a forger $\mathcal{B}$ capable of breaking the pqRSA encryption scheme. The forger $\mathcal{B}$ receives as input a public key (N, e) and tries to invert $x = f^{-1}(y)$, where f is the pqRSA function defined by N and e. The adversary $\mathcal{B}$ starts running the proxy P for this public key and a randomly chosen number d_1 given as a secret key. When the proxy P makes the i-th hash query, the adversary looks to see if the message m_i was already asked. If not, it picks a random x_i, sets $h(m_i) = x_i^e$ with probability p and $h(m_i) = y * x_i^e$ with probability $1 - p$. If the proxy P makes a query to $FSig$ for a message m_i, the adversary returns x_i if m_i was asked before. Otherwise it fails. Eventually, P outputs a correct unidirectional RSA-Hash signature (m, s) for a brand new message m. If the message m was not hashed before, the adversary computes its hash value. If $h(m) = y * x_i^e$, then the adversary returns $y^d = s/x_i^e$ as $x = f^{-1}(y)$. Otherwise, it fails.

2. Let's assume that F can break the unidirectional pqPFDH scheme. This means that $|Success_{F,\mathcal{S}}(1^k)|$ is not negligible. Based on the user F, we build a forger $\mathcal{B}$ capable of breaking the pqRSA encryption scheme. The forger $\mathcal{B}$ receives as input a public key (N, e) and tries to invert $x = f^{-1}(y)$, where f is the pqRSA function defined by N and e. The adversary $\mathcal{B}$ starts running the user F for this public key and a randomly chosen number d_2 given as a secret key. When the user F makes the i-th hash query, the adversary looks to see if the message mi was already asked. If not, it picks a random x_i, sets $h(m_i) = x_i^e$ with probability p and $h(m_i) = y * x_i^e$ with probability $1 - p$. If the user F makes a query to $PSig$ for a message m_i, the adversary returns x_i if m_i was asked before. Otherwise it fails. Eventually, F outputs a correct unidirectional RSA-Hash signature (m, s) for a brand new message m. If the message m was not hashed before, the adversary versary computes its has value. If $h(m) = y * x_i^e$, then the adversary returns $y^d = s/x_i^e$ as the $x = f^{-1}(y)$. Otherwise, it fails.
3. The security proof against a user U is similar to the two previous parts.

The security reduction for the pqPFDH signature scheme is tight because it is equivalent to those of pqRSA. With a random of 5 bits, we have $\varepsilon_{\mathcal{R}} = 0.96875\ \varepsilon_{\mathcal{A}}$, where $\varepsilon_{\mathcal{R}}$ is the success probability of an algorithm $\mathcal{R}$ that can break RSA by using an attacker $\mathcal{A}$ that breaks pqPFDH with probability $\varepsilon_{\mathcal{A}}$. But for RSA-PFDH, it is necessary to use random with size greater than $\log_2 q_{sig} + 8\ (\geq 38)$, where q_{sig} is the number of signatures queries, in order to have $\varepsilon_{\mathcal{R}} = 1.04\ \varepsilon_{\mathcal{A}}$. We also remark that the success probability of the simulation is independent of the number of signing and hashing oracles queries. About the established unidirectional pqPFDH, it is proven to be unforgeable against chosen message attack since the security of the corresponding standard scheme is proven. So, the security is also guaranteed against the proxy P and the user F that take shared secret keys but also for any other user U.

6 Conclusion

In this work, we described the unidirectional version of the signature scheme called post-quantum Probabilistic Full Domain Hash (pqPFDH) which is a variant of the RSA-FDH signature scheme but with an approach that allows it to be used in a quantum area. It is proven secure against chosen message attack in the random oracle model. The security reduction has a better security proof than PFDH because much shorter random salt is sufficient to achieve the same security level. We use the formalized definition of unidirectional proxy function for signature to establish unidirectional version of the pqPFDH signature scheme. This allows one user to delegate his signature right to another two users (the proxy P and the user F) that in turn will be able to sign on behalf of him. The security of this last construction is relatively linked to that of the former, which prove that the unidirectional pqPFDH is secure against P, F, and U.

For previous work, it is important to establish the bidirectional version of the pqPFDH signature Scheme using proxy function for both signature and encryption.

It is also a good idea to prove that the pqPFDH scheme has a history-free reduction which guarantees its security in the quantum accessible random oracle model.

References

1. Blazz, M., Strauss, M.: Atomic proxy cryptography. In: Proceedings of EuroCrypt '97 (1998)
2. Blazz, M., et al.: Divertible protocols and atomic proxy cryptography. In: International Conference on the Theory and Applications of Cryptographic Techniques, Advances in Cryptology—EUROCRYPT'98, pp. 127–144
3. Bégueline, S., Grégoire, B.: Formally certifying the security of digital signature schemes. In: 30th IEEE Symposium in Security and Privacy (2009)
4. Bellare, M., Rogaway, P.: Random oracles are practical: a paradigm for designing efficient protocols. In: Proceedings of the First Annual Conference on Computer and Communications Security. ACM (1993)
5. Bellare, M., Rogaway, P.: The Exact security of digital signatures: how to sign with RSA and Rabin. In: Proceedings of Eurocrypt 1996, LNCS, vol. 1070, pp. 399–416. Springer (1996)
6. Bellare, M., Sandhu, R.: The Security of Practical Two-Party RSA Signature Schemes, Cryptology ePrint Archive: Report 2001/060
7. Boneh, D., et al.: Random oracles in a quantum world. In: International conference on the theory and Application of Cryptology and Information Security. ASIACRYPT 2011, pp. 41–69. Springer
8. Cachin, C., et al.: Computationally private information retrieval with polylogarithmic communication. In: Advances in Cryptology EUROCRYPT 99. Lecture Notes in Computer Science, vol. 1592, pp. 402–414. Springer, Berlin, Heidelberg (1999)
9. Canetti, R., Goldreich, O., Halevi, S.: The random oracle methodology, revisited. J. ACM **51**(4), 557–594 (2004)
10. Coron, J.S.: On the exact security of Full Domain Hash. In: Proceedings of Crypto's 2000, LNCS, vol. 1880, pp. 229–235. Springer (2000)
11. Coron, J.S.: Optimal security proofs for PSS and other signature schemes. In: Proceedings of Eurocrypt'02, LNCS, vol. 2332, pp. 272–287. Springer (2002)
12. Dodis, Y., Reyzin, L.: On the power of claw-free permutations. In: Security in Communication Networks (SCN 2002). Lecture Notes in Computer Science, vol. 2576, pp. 55–73. Springer (2003)
13. Dodis, Y., Oliveira, R., Pietrzak, K.: On the generic insecurity of full-domain hash. In: Advances in Cryptology-CRYPTO 2005: 25th Annual International Cryptology Conference, Santa Barbara, California, USA, August 14–18, 2005, Proceedings
14. Goldwasser, S., Micali, S., Rivest, R.: A digital signature scheme secure against adaptive chosen-message attacks. SIAM J. Comput.**17**(2), 281–308 (1988)
15. Hue, Y., et al.: Easy simple factoring-based digital signature scheme. In: 15th International Conference for Internet Technology and Secured Transactions (ICITST). IEEE (2020)
16. Ivan, A., Dodis, Y.: Proxy cryptography revisited. In: Proceeding of the 10th Annual Network and Distributed System Security Symposium-NDSS'03
17. Katz, J., Wang, N.: Efficiency improvements for signature schemes with tight security reductions. In: Conference on Computer and Communications Security (2003). www.dl.acm.org
18. Kiltz, E., Kakvi, S.: Optimal security proofs for full domain hash, revisited. J. Cryptol. **31**, 276–306 (2018). https://doi.org/10.1007/s00145-017-9257-9
19. Kim, H., et al.: Secret computation with secrets for mobile agent using one-time proxy signature. In: The 2001 Symposium on Cryptography and Information Security (SCIS 2001), Oiso, Japan, January 23–26 (2001)

20. MacKenzie, P., Reiter, K.: Two-party generation of DSA signatures. In: Advances in Cryptology—CRYPTO 2001. Lecture Notes in Computer Science, vol. 2139 (2001)
21. MacKenzie, P., Reiter, K.: Networked cryptographic devices resilient to capture. In: Eighth ACM Conference on Computer and Communications Security (CCS-8) (2001)
22. Mambo, M., Usuda, K., Okamoto, E.: Proxy signatures for delegating sign operation. In: Proceedings of the 3rd ACM Conference on Computer and Communications Security (CCS96), pp. 48–57. ACM Press (1996)
23. Mbaye, M.L., Sow, D, Sow, D: The psot-quantum probabilistic full domain hash. In: Proceedings of the 8th international Cryptology and Information Security Conference 2022, CRYPTOLOGY22, Putrajaya, 26–28 July 2022, pp. 50–68. Code 182061, indexed by Scopus (2022)
24. Pointcheval, D., Stern, J.: Security proofs for signature schemes. In: Maurer, U. (ed.) Advances in Cryptology Proceedings of EUROCRYPT '96 (May 12–16, 1996, Zaragoza, Spain). LNCS, vol. 1070, pp. 387–398. Springer (1996)
25. Rivest, R.: A method for obtaining digital signatures and public key cryptosystems. CACM *21*, (1978)
26. Vassilev, A., et al.: NIST roadmap toward criteria for threshold schemes for cryptographic primitives (2020). https://doi.org/10.6028/NIST.IR.8214A
27. Zhandry, M.: Secure identity-based encryption in the quantum random oracle model. Int. J. Quantum Inf. **13**(04), 1550014 (2015)

Modeling Malaria Transmission by the Movement of Mosquitoes from Hot Spot Area to Areas of Low Transmission

Moussa Kane, Karim Konate, Ousmane Sy, El Hadji Amadou Niang, Cheikh Gaye, Mamadou Saliou Diallo, Ibrahima Ndiaye, Yacine Aw, Almamy Youssouf Ly, Léontine Ndogou Bakhoum, Khady Ndiaye, Edouard Guédj Tine, Moussa Diop, Lucien Gning, Mor Absa Loum, Mamadou Bousso, Lassana Konate, Ousmane Faye, Oumar Gaye, and Jean Louis Ndiaye

Abstract Our study proposes an extended SIR model that explicitly incorporates mosquito mobility between high-transmission (hot spot) and low-transmission areas to explain malaria persistence or resurgence. We establish a local threshold governed by the basic reproduction number: below this threshold, introductions die out; above it, an endemic equilibrium exists and is locally asymptotically stable. Vector mobility

M. Kane (✉) · K. Konate · O. Sy · E. H. A. Niang · C. Gaye · M. S. Diallo · I. Ndiaye · Y. Aw · E. G. Tine · M. Diop · L. Konate · O. Faye · O. Gaye
Cheikh Anta Diop University, Dakar, Senegal
e-mail: moussa4.kane@ucad.edu.sn

K. Konate
e-mail: karim.konate@ucad.edu.sn

O. Sy
e-mail: syousmane7@gmail.com

E. H. A. Niang
e-mail: elhadjiamadou.niang@ucad.edu.sn

C. Gaye
e-mail: gaye80070@gmail.com

M. S. Diallo
e-mail: mamadousalioudiallo378@gmail.com

I. Ndiaye
e-mail: ndiayebro23@gmail.com

Y. Aw
e-mail: yacineaw89@gmail.com

E. G. Tine
e-mail: edouguedjtine@gmail.com

L. Konate
e-mail: lassana.konate@ucad.edu.sn

© The Author(s), under exclusive license to Springer Nature Switzerland AG 2026

C. T. Gueye et al. (eds.), *Mathematics, Computer Science, Cybersecurity and Artificial Intelligence*, Springer Proceedings in Mathematics & Statistics 536,
https://doi.org/10.1007/978-3-032-17666-0_7

plays an ambivalent role: it reduces the residence of susceptible mosquitoes in the recipient area (a structural effect that lowers the threshold), but if a flux of infected mosquitoes from a hot spot persists, the same mobility acts as an external forcing capable of maintaining low-level endemicity, especially near the threshold. We derive closed-form expressions for the endemic equilibrium and prove its stability, then numerically confirm the expected regimes (extinction, endemicity, seasonal resurgence under importation). Operationally, our results argue for coupled strategies: lowering local transmission potential (reducing mosquito–human contact, shortening vector survival, rapid case management) and controlling "entomological frontiers" to limit inflows from hot spots, with seasonally tuned interventions and sensitive surveillance. This framework thus provides an analytical and practical basis for guiding control efforts in low-transmission settings.

Keywords Malaria · Mosquito mobility · Extended SIR model · Basic reproduction number (R0) · Endemic equilibrium · Disease · Free equilibrium (DFE) · Stability analysis (Jacobian)

O. Faye
e-mail: jogomaye@yahoo.fr

M. Kane · K. Konate · O. Gaye
Department of Mathematics and Computer Science, Faculty of Science and Technology, UCAD, Dakar, Senegal

O. Sy · L. Gning · M. A. Loum · M. Bousso
Malaria Research Capacity Development, MARCAD, Dakar, Senegal
e-mail: lucien.gning@univ-thies.sn

M. A. Loum
e-mail: morabsa.loum@univ-thies.sn

M. Bousso
e-mail: mbousso@univ-thies.sn

E. H. A. Niang · O. Faye · J. L. Ndiaye
LEVP, UCAD, Laboratory of Vector and Parasitic Ecology, Dakar, Senegal

A. Y. Ly · L. N. Bakhoum · K. Ndiaye · E. G. Tine · M. Diop
University Iba Der Thiam of Thies, UIDT, Thies, Senegal
e-mail: ayoussouf.ly@univ-thies.sn

L. N. Bakhoum
e-mail: lndogou.bakhoum@univ-thies.sn

K. Ndiaye
e-mail: khady.ndiaye@aims-senegal.org

1 Introduction

Vector-borne diseases, particularly those transmitted by mosquitoes, represent a major threat to global public health. Understanding transmission dynamics and the interactions between human and vector populations is essential for designing effective prevention and control strategies. Among the mathematical models developed to analyze the spread of infectious diseases, the SIR (Susceptible–Infected–Recovered) model remains one of the most fundamental and widely applied.

However, to adequately capture the complexity of vector-borne diseases, it is necessary to move beyond this classical framework. In this study, we extend the SIR model by explicitly incorporating a mosquito population, thereby improving the representation of epidemiological dynamics for diseases such as malaria, dengue, and Zika virus. This enriched model considers susceptibility, infection, and recovery in humans, as well as susceptibility and infection in mosquitoes.

Beyond this structural extension, a key question remains insufficiently explored: the role of mosquito movement from high-transmission areas to low-transmission areas in the spread and persistence of malaria. This issue, both scientific and operational, poses a crucial challenge: to what extent can vector mobility compromise control achievements in areas that were previously unaffected or characterized by low transmission?

To address this question, our approach relies on a rigorous analysis of the stability of the system's endemic equilibrium. This equilibrium corresponds to a situation in which the disease persists in a stable manner within human and vector populations, without uncontrolled outbreaks or complete extinction. By examining the Jacobian matrix at this equilibrium, we assess whether small perturbations—particularly those induced by mosquito movements—will drive the system back toward equilibrium or, conversely, trigger a profound shift in epidemiological dynamics.

Thus, the primary objective of this research is clear: to model and analyze the impact of vector mobility between high- and low-transmission areas of malaria. We hypothesize that mosquito dispersal is a central mechanism of epidemic resurgence, capable of sustaining transmission in areas where the disease might otherwise disappear in the absence of vector migration. Testing this hypothesis not only refines our understanding of the spatio-temporal dynamics of malaria but also provides public health decision-makers with practical evidence to guide the planning of interventions tailored to local transmission contexts.

To validate our theoretical results, we also develop a numerical simulation in R. This simulation allows us to visualize population dynamics over time, explore different scenarios of vector mobility, and confirm or refute the stability of the endemic equilibrium.

2 State of the Art

Mathematical modeling of infectious diseases is a well-established field of research that has considerably advanced our understanding of epidemiological dynamics. The pioneering works of Ross [1] and later Macdonald [2] laid the foundations of theoretical malaria epidemiology by highlighting the central role of entomological parameters such as biting rate, transmission probability, and vector longevity. These approaches led to the formalization of the basic reproduction number and to the first structured strategies for disease control.

Subsequent developments in epidemiological theory, particularly by Diekmann and Heesterbeek [3], provided more rigorous analytical tools for studying equilibrium stability and eradication thresholds. In the same line, Chitnis et al. [4] applied these methods to vector-borne models, showing how reducing vector populations could alter system stability and potentially enable elimination. In parallel, compartmental models have been enriched to include host- and vector-specific parameters. Smith et al. [5], for instance, revisited Ross–Macdonald-type models to integrate biting and recovery rates, while Ferguson et al. [6] developed stochastic models to evaluate the impact of interventions such as insecticide-treated nets or indoor residual spraying. These approaches demonstrated the importance of incorporating uncertainty and parameter variability to guide public health policies.

More recently, research has focused on understanding the role of mosquito movement in transmission dynamics. Mark–release–recapture (MRR) studies have shown that local dispersal of *Aedes aegypti* generally remains limited to a few hundred meters, with a median of around 90 m, strongly influenced by urban structure and wind conditions [7, 8]. These findings underscore that transmission connectivity operates at a microgeographic scale and must be incorporated into spatio-temporal models. At a larger scale, major discoveries have highlighted the importance of wind-borne migrations. Huestis et al. [9] demonstrated that mosquitoes can travel hundreds of kilometers carried by high-altitude winds in the Sahel, establishing entomological connections between distant regions. This long-distance migration redefines our understanding of transmission boundaries and explains the reappearance of outbreaks despite unfavorable local conditions.

In addition, the recent invasion of *Anopheles stephensi* in Africa illustrates vector dynamics in a context of rapid urbanization. Using ecological niche models, Sinka et al. [10] showed that many African urban areas are at high risk of establishment of this vector. This prediction was confirmed by epidemiological evidence from Emiru et al. [11], who associated the presence of *An. stephensi* with urban malaria outbreaks. More recently, Taylor et al. [12] synthesized lessons from Asia and identified specific African challenges, including water management and rapid urban expansion.

In response to these new data, modeling has incorporated explicit spatial dimensions. Kissler et al. [13] conducted a systematic review of spatial connectivity models, highlighting differences between gravity, radiation, and metapopulation approaches.

Similarly, Citron et al. [14] compared the performance of these models and demonstrated their impact on epidemiological predictions, particularly for targeting high-risk areas. Finally, reaction–diffusion approaches have been applied to better understand the spatial propagation of infection fronts, as shown by Wang et al. [15], who integrated seasonality and drug resistance into their analysis.

In summary, both classical and recent studies show that the transmission dynamics of vector-borne diseases cannot be understood without considering vector movement—whether through local dispersal, long-distance migrations, or the invasion of new urban environments. The integration of these spatial dynamics into epidemiological models is now essential for anticipating and controlling the spread of malaria and other vector-borne diseases.

Limitations of Existing Models and Positioning of Our Study

While classical SIR models remain effective for describing the dynamics of many infectious diseases, they exhibit a major limitation when applied to vector-borne infections: they do not account for vector movement. Yet mosquito mobility plays a decisive role in the spread of malaria, dengue, and Zika virus. By neglecting these dynamics, traditional models oversimplify interactions between vectors and their environment and overlook essential drivers of transmission.

This limitation also reduces the effectiveness of geographically targeted interventions. For instance, insecticide spraying or bed net distribution in a specific area may prove insufficient if infected mosquitoes migrate from untreated neighboring zones, leading to unexpected persistence or even resurgence of disease. A fine understanding of vector movement and its impact on transmission is thus indispensable for optimizing control strategies.

Despite notable progress, several research gaps remain. First, empirical studies are often fragmented across spatial scales: some focus on short-range dispersal (a few hundred meters), while others document windborne migration over hundreds of kilometers. Rarely are these two scales—micro-dispersion and regional migration—integrated into a unified framework, which limits comprehensive understanding of vector connectivity. Second, African contextualization is still insufficient. While the emergence of *Anopheles stephensi* has been the subject of recent studies, most address urban invasion dynamics rather than heterogeneous transmission conditions such as those observed in Senegal.

Furthermore, existing models remain overly simplified. Most rely on homogeneous systems of ordinary differential equations that ignore vector flows between regions. Even more advanced spatial models, based on metapopulations or reaction–diffusion equations, remain largely theoretical and rarely calibrated with local empirical data. Finally, the literature addresses only superficially the critical issue of malaria reintroduction. Although many studies analyze persistence in endemic settings, few investigate the risk of resurgence in low-transmission or near-elimination areas—a challenge that is particularly relevant for West Africa.

In light of these limitations, our study provides an original contribution. We propose an enriched compartmental model that extends the SIR framework by explicitly incorporating mosquito mobility between high- and low-transmission zones. This approach offers a more realistic representation of malaria reintroduction dynamics and evaluates the impact of vector movement on equilibrium stability. Our analysis combines mathematical tools, stability analysis and calculation of the basic reproduction number (R0), with numerical simulations implemented in R (using *deSolve* and *ggplot2*), ensuring both theoretical and practical validation.

Crucially, our study is contextually relevant to Senegal, a country where high-transmission areas in the south and southeast coexist with low-transmission areas in the center and north. By integrating these regional contrasts, we explore realistic scenarios of resurgence driven by vector flows. Our findings aim to inform public health decision-makers about the need to strengthen surveillance of "entomological frontiers," to coordinate interventions across scales, and to design targeted measures that prevent malaria reintroduction in vulnerable areas.

In sum, our approach addresses a critical gap in the literature by combining mathematical stability analysis with a model explicitly tailored to the West African context, providing a robust tool to anticipate and counter the risks of malaria reintroduction linked to mosquito movement.

3 Methodology

To overcome the limitations of classical SIR models and propose a more realistic representation of malaria dynamics, our study follows a three-step methodology combining mathematical modeling, theoretical analysis, and numerical validation through simulation.

(a) Definition of the Mathematical Model

We develop an extended compartmental model that explicitly incorporates the mosquito population and their movements between zones with heterogeneous transmission intensity. This model accounts for:

- The dynamics of human and mosquito populations, including processes of birth, mortality, infection, recovery, and loss of immunity.
- Mosquito mobility, represented by migration terms between high- and low-transmission zones, and influenced by environmental factors such as seasonality, climatic conditions, and urbanization.
- Human–vector interactions through transmission parameters (inoculation probability, biting rate, infectious period), capturing the complexity of epidemiological exchanges.

(b) **Theoretical Analysis and Stability Study**

Based on this model, we conduct a rigorous mathematical analysis to characterize its dynamic properties:

- Identification of equilibrium points, including the disease-free equilibrium and the endemic equilibrium.

- Examination of local and global stability using the Jacobian matrix and eigenvalue analysis.
- Calculation of the basic reproduction number (R0), modified to include vector mobility, in order to determine the conditions under which the disease goes extinct, persists, or is reintroduced.

(c) **Numerical Implementation and Simulations**

Finally, to confront the analytical results with realistic scenarios, we implement the model in **R** using the *deSolve* package (for solving differential equations) and *ggplot2* (for visualization). The numerical simulations allow us to:

- Explore the impact of key parameters on transmission, particularly the role of vector mobility.
- Test different epidemiological scenarios, including variations in human and vector population densities, the introduction of public health interventions (e.g., insecticide-treated nets, indoor residual spraying), and climatic fluctuations influencing mosquito dispersal.
- Compare the resulting dynamics across high- and low-transmission zones to assess the risk of malaria reintroduction and to identify the conditions favoring resurgence.

4 Mathematical Analysis of the Model

4.1 Model Formulation

The diagram in Fig. 1 illustrates the transmission dynamics of malaria between humans and mosquitoes, with vectors moving from high-transmission areas to low-transmission areas.

We now formulate the differential equations based on these definitions.

$$\begin{cases} \frac{dS_h}{dt} = \Lambda_h N_h - a\beta I_m \frac{S_h}{N_h} - \mu_h S_h + \nu R_h (1) \\ \frac{dI_h}{dt} = a\beta I_m \frac{S_h}{N_h} - (\mu_h + \gamma + b) I_h (2) \\ \frac{dR_h}{dt} = \gamma I_h - (\mu_h + \nu) R_h (3) \\ \frac{dS_m}{dt} = \mu N_m - a\beta_m I_h \frac{S_m}{N_h} - \mu_m S_m - m S_m (4) \\ \frac{dI_m}{dt} = a\beta_m I_h \frac{S_m}{N_h} - \mu_m I_m + m I_{m_h} (5) \end{cases}$$

Explanation of the Differential Equations

- **Equation for the susceptible human population** (S_h):

$$\frac{dS_h}{dt} = \Lambda_h N_h - a\beta I_m \frac{S_h}{N_h} - \mu_h S_h + \nu R_h$$

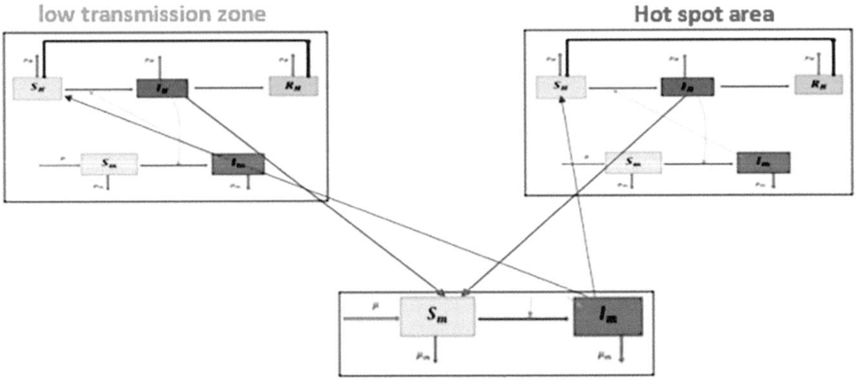

Fig. 1 Diagram of malaria importation by mosquitoes from high-transmission areas to low-transmission areas. S_h: Susceptible humans, I_h: Infected humans, R_h: Recovered humans, S_m: Susceptible mosquitoes, I_m: Infected mosquitoes, N_h Total human population, N_m: Total mosquito population, I_{mh}: Infected mosquitoes in a high-transmission area, S_{ml}: Susceptible mosquitoes in a low-transmission area, $\wedge_h$: Birth or immigration rate of susceptible humans, α: Number of bites, β: Disease transmission rate to humans, b: Human mortality rate due to the disease, γ: Human recovery rate, μ: Birth or immigration rate of susceptible mosquitoes, μ_h: Natural mortality rate of humans, μ_m: Natural mortality rate of mosquitoes, β_m: Disease transmission rate to mosquitoes, ν: Immunity loss rate, m: Mosquito movement rate between two areas, indicating the proportion of mosquitoes moving from a high-transmission area to a low-transmission area

- $\Lambda_h N_h$: Birth or immigration rate of susceptible humans.
- $-a\beta I_m \frac{S_h}{N_h}$: Infection of susceptible humans after being bitten by an infected mosquito (I_m); β is the transmission rate, and a represents the number of bites per mosquito.
- $-\mu_h S_h$: Natural death of susceptible humans.
- νR_h : Loss of immunity in recovered humans, who return to the susceptible class.

• **Equation for the infected human population** (I_h):

$$\frac{dI_h}{dt} = a\beta I_m \frac{S_h}{N_h} - (\mu_h + \gamma + b)I_h$$

- $a\beta I_m \frac{S_h}{N_h}$: Infection of susceptible humans after being bitten by an infected mosquito.
- $-(\mu_h + \gamma + b)I_h$: Combined effects of:

  $-\mu_h I_h$: Natural death of infected humans.
  $-\gamma I_h$:: Recovery of infected humans, who move into the recovered class (R_h).
  $-bI_h$: Disease-induced mortality.

- **Equation for the recovered human population** (R_h):

$$\frac{dR_h}{dt} = \gamma I_h - (\mu_h + \nu)R_h$$

  - γI_h: Recovery of infected humans, who become recovered.
  - $-(\mu_h + \nu)R_h$: Combined effects of:

    $-\mu_h R_h$: Natural death of recovered humans.
    $-\nu R_h$: Loss of immunity, returning recovered individuals to the susceptible class (S_h).

- **Equation for the susceptible mosquito population** (S_m):

$$\frac{dS_m}{dt} = \mu N_m - a\beta_m I_h \frac{S_m}{N_h} - \mu_m S_m - mS_m$$

  - μN_m: Birth or immigration of susceptible mosquitoes.
  - $-a\beta_m I_h \frac{S_m}{N_h}$: Infection of susceptible mosquitoes after biting an infected human.
  - $-\mu_m S_m$: Natural death of susceptible mosquitoes.
  - $-mS_m$: Proportion mm of susceptible mosquitoes moving to another area.

- **Equation for the infected mosquito population** ($\boldsymbol{I_m}$):

$$\frac{dI_m}{dt} = a\beta_m I_h \frac{S_m}{N_h} - \mu_m I_m + mI_{m_h}$$

  - $a\beta_m I_h \frac{S_m}{N_h}$: Infection of susceptible mosquitoes after biting an infected human.
  - $-\mu_m I_m$: Natural death of infected mosquitoes.
  - $+mI_{m_h}$: Migration of infected mosquitoes from a high-transmission area into a low-transmission area.

This model extends the classical SIR framework by incorporating mosquito mobility between zones, thereby enabling the study of how vector migration influences malaria transmission. This is essential for understanding transmission hotspots and refining control strategies such as insecticide spraying or the distribution of bed nets in targeted areas. The inclusion of the movement parameter m provides a more realistic representation of spatial dynamics, a crucial improvement over classical models that assume homogeneous and static populations.

4.2 The Jacobian Matrix at the Disease-Free Equilibrium

The Jacobian matrix of the system is obtained by linearizing the model equations around the disease-free equilibrium. The disease-free equilibrium (E_0) is defined as

$$E_0 = (S_h^*, I_h^* = 0, R_h^*, S_m^*, I_m^* = 0)$$

where S_h^* and S_m^* represent the equilibrium values of the susceptible human and mosquito populations in the absence of disease.

Pour les moustiques susceptibles:

$$\frac{dS_m}{dt} = \mu N_m - a\beta_m I_h \frac{S_m}{N_h} - \mu_m S_m - mS_m$$

At the DFE (disease-free equilibrium), $I_h = 0$ er $I_m = 0$.

The infection term vanishes, and the equilibrium satisfies $\frac{dS_m}{dt} = 0 \Rightarrow$ $\boxed{S_m^* = \frac{\mu}{\mu_m + m} N_m}$.

In the recipient area, the resident size of susceptible mosquitoes is a fraction of the reservoir N_m, reduced by natural mortality μ_m outward migration m.

Considering that the human population is constant, the Jacobian matrix at the disease-free equilibrium is

$$J(E_0) = \begin{bmatrix} -\mu_h & 0 & \nu & 0 & -a\beta\frac{S_h^*}{N_h} \\ 0 & -(\mu_h + \gamma + b) & 0 & 0 & a\beta\frac{S_h^*}{N_h} \\ 0 & \gamma & -(\mu_h + \nu) & 0 & 0 \\ 0 & -a\beta_m\frac{S_m^*}{N_h} & 0 & -\mu_m - m & 0 \\ 0 & a\beta_m\frac{S_m^*}{N_h} & 0 & 0 & -\mu_m \end{bmatrix}$$

Since $S_h^* = N_h$, then $\frac{S_h^*}{N_h} = 1$, we have

$$J(E_0) = \begin{bmatrix} -\mu_h & 0 & \nu & 0 & -a\beta \\ 0 & -(\mu_h + \gamma + b) & 0 & 0 & a\beta \\ 0 & \gamma & -(\mu_h + \nu) & 0 & 0 \\ 0 & -a\beta_m\frac{S_m^*}{N_h} & 0 & -\mu_m - m & 0 \\ 0 & a\beta_m\frac{S_m^*}{N_h} & 0 & 0 & -\mu_m \end{bmatrix}$$

The characteristic matrix:

$$p(\lambda) = \lambda I - J(E_0) = \begin{bmatrix} \lambda + \mu_h & 0 & -\nu & 0 & a\beta \\ 0 & \lambda + \mu_h + \gamma + b & 0 & 0 & -a\beta \\ 0 & -\gamma & \lambda + \mu_h + \nu) & 0 & 0 \\ 0 & a\beta_m \frac{S_m^*}{N_h} & 0 & \lambda + \mu_m + m & 0 \\ 0 & -a\beta_m \frac{S_m^*}{N_h} & 0 & 0 & \lambda + \mu_m \end{bmatrix}$$

It is observed that the first column contains only one non-zero term $(\lambda + \mu_h)$, we therefore develop according to the first column $\det(\lambda I - J)$

$$p(\lambda) = \det(\lambda I - J) = (\lambda + \mu_h) detM_{1,1}$$

where $M_{1,1}$ is the 4×4 submatrix obtained by deleting the first row and the first column.

Expansion of $detM_{1,1}$
We have

$$\boldsymbol{M}_{1,1} = \begin{bmatrix} +\mu_h + \gamma + b & 0 & 0 & -a\beta \\ -\gamma & \lambda + \mu_h + \nu & 0 & 0 \\ a\beta_m \frac{S_m^*}{N_h} & 0 & \lambda + \mu_m + m & 0 \\ -a\beta_m \frac{S_m^*}{N_h} & 0 & 0 & \lambda + \mu_m \end{bmatrix}$$

The second column of $\boldsymbol{M}_{1,1}$ has only one non-zero term $(\lambda + \mu_h + \nu)$ in position 2×2. Expanding again:

$$det\, M_{1,1} = (\lambda + \mu_h + \nu)\, det\, M_{(1,1),(2,2)}$$

where $M_{(1,1),(2,2)}$ is the 3×3 submatrix obtained by deleting the second row and the second column:

$$M_{(1,1),(2,2)} = \begin{pmatrix} \lambda + \mu_h + \gamma + b & 0 & -a\beta \\ a\beta_m \frac{S_m^*}{N_h} & \lambda + \mu_m + m & 0 \\ -a\beta_m \frac{S_m^*}{N_h} & 0 & \lambda + \mu_m \end{pmatrix}$$

In this 3×3, the second column has only one non-zero term $(\lambda + \mu_m + m)$ (in $(2, 2)$). Expanding once more:

$$det\, M_{(1,1),(2,2)} = (\lambda + \mu_m + m) detC$$

with the infectious block 2×2:

$$C = \begin{pmatrix} \lambda + \mu_h + \gamma + b & -a\beta \\ -a\beta_m \frac{S_m^*}{N_h} & \lambda + \mu_m \end{pmatrix}$$

Its determinant is

$$\det C = (\lambda + \mu_h + \gamma + b)(\lambda + \mu_m) - a^2 \beta \beta_m \frac{S_m^*}{N_h}$$

Piecing everything together, we obtain the complete factorization:

$$\boxed{\mathrm{p}(\lambda) = (\lambda + \mu_h)(\lambda + \mu_h + \nu)(\lambda + \mu_m + m)\left[\underbrace{(\lambda + \mu_h + \gamma + b)(\lambda + \mu_m) - a^2 \beta \beta_m \frac{S_m^*}{N_h}}_{\text{infectious block}}\right]}$$

- **Eigenvalues and stability condition**
  - Three eigenvalues are immediate and strictly negative:

$$\lambda_1 = -\mu_h, \lambda_2 = -(\mu_h + \nu), \lambda_3 = -(\mu_m + m)$$

- The two remaining ones are the roots of the infectious quadratic:

$$q(\lambda) = \lambda^2 + \lambda((\mu_h + \gamma + b) + \mu_m) + ((\mu_h + \gamma + b)\mu_m - a^2 \beta \beta_m \frac{S_m^*}{N_h})$$

By Routh–Hurwitz (for $\lambda^2 + A\lambda + C$).

- Stability $\Leftrightarrow A > 0 and C > 0$

Here $A = (\mu_h + \gamma + b) + \mu_m > 0$
The crucial condition is

$$C > 0 \Leftrightarrow \frac{a^2 \beta \beta_m}{(\mu_h + \gamma + b)\mu_m} \frac{S_m^*}{N_h} < 1$$

Defining

$$R_0^2 = \frac{a^2 \beta \beta_m}{(\mu_h + \gamma + b)\mu_m} \frac{S_m^*}{N_h}$$

we obtain the threshold condition:

E_0 is locally asymptotically stable if $R_0 < 1$, and unstable if $R_0 > 1$
With $S_m^* = \frac{\mu}{\mu_m + m} N_m$,

$$\boxed{R_0 = \sqrt{\frac{a^2\beta\beta_m}{(\mu_h+\gamma+b)\mu_m}\frac{N_m}{N_h}\frac{\mu}{\mu_m+m}}}$$

- a: biting frequency; $\beta\beta_m$: transmission probabilities (mosquito $\rightarrow$ human, human $\rightarrow$ mosquito).
- $\mu_h+\gamma+b$: exit rate from I_h (natural death, recovery, disease-induced death); μ_m: mosquito mortality.
- $\frac{N_m}{N_h}$: relative vector density; $\frac{\mu}{\mu_m+m}$: "resident" availability of susceptible mosquitoes

- **What does the model imply for a "low-transmission" zone?**
  - **Without importation (no flux from the hot spot)**

    Local threshold
    A **low-transmission** zone satisfies $R_0 < 1$: any small introduction of infection decays (negative eigenvalues) $\rightarrow$ spontaneous return to the DFE.

  - **With vector importation (inflow + mI_{mh})**

Even if $R_0 < 1$, an exogenous forcing term can maintain a non-zero infection level.
Linearizing around the DFE (where $S_h \approx N_h$, $S_m \approx S_m^*$):

$$a\beta I_m \frac{S_h}{N_h} \approx (\mu_h+\gamma+b)I_h$$

$$a\beta_m I_h \frac{S_m}{N_h} \approx \mu_m I_m - mI_{mh}$$

Combining these yields the first-order forced equilibrium:

$$I_h^* \approx \frac{ma\beta}{\mu_m(\mu_h+\gamma+b)\left(1-R_0^2\right)} I_{mh}, \; I_m^* \approx \frac{\mu_h+\gamma+b}{a\beta} I_h^*$$

Interpretation

- The closer R_0 is to **1**, the larger the gain $\frac{1}{1-R_0^2}$ $\rightarrow$ the zone becomes hyper-sensitive to even small.
- If the inflow **stops**, the system relaxes back to the DFE (exponential decay).

- **Effect of the mobility parameter m in the recipient zone**
  - **"Protective" effect (via R_0)**

    m reduces the resident pool of susceptible mosquitoes:

$S_m^* = \frac{\mu}{\mu_m+m} N_m \;\Rightarrow R_0^2$ is proportional to $\frac{1}{\mu_m+m} \Leftrightarrow R_0^2 \propto \frac{1}{\mu_m+m}$
Log-sensitivity: $\frac{\partial lnR_0}{\partial m} = -\frac{1}{2(\mu_m+m)} < 0$
Thus, increasing m (more outward movement from the recipient zone) lowers R_0 $\rightarrow$ self-sustained transmission becomes harder.

- **"Adverse" effect (via importation)**

The same m amplifies the inflow of infected mosquitoes mI_{mh}
For $R_0 < 1$, the forced equilibrium scales as

$$I_h^* \propto \frac{m}{1 - R_0^2(m)} I_{mh}$$

If R_0 is well below 1, the denominator dominates $\rightarrow$ the inflow effect is contained. If $R_0 \rightarrow 1^-$, the denominator collapses $\rightarrow$ large surges even for modest inflows. **Consequence:** the overall response to m can be non-monotonic (an "intermediate" m may maximize incidence if R_0 is near 1).

- **Which lever to prioritize?**
  - If $R_0 \ll 1$: incidence is driven by the incoming flux $\rightarrow$ priority is to reduce $\boldsymbol{I_{mh}}$ (act at the hot spot).
  - If $R_0 \approx 1$: priority is to push R_0 locally ($a \searrow, \beta \searrow, \beta_m \searrow$, $\mu_m \nearrow$); otherwise, each importation can rekindle transmission.

- **Canonical scenarios for a low-transmission zone**
  - $R_0 < 1$, no importation ($\boldsymbol{I_{mh}} = 0$)

    Trajectory: every introduction dies out (exponential decay)
    Action: routine surveillance; no heavy intervention needed.

  - $-R_0 < 1$, weak seasonal importations

    Trajectory: transient waves aligned with seasons (winds, rains). Amplitude increases as $R_0 \rightarrow 1^-$
    Action: seasonal boosts (e.g., pre-season IRS/LLIN), early-warning triggers.

  - $R_0 < 1$, sustained importation (constant flux)

    Trajectory: forced endemicity (low but persistent prevalence).
    Action: dual lever source + recipient**:**
    Source: reduce $\boldsymbol{I_{mh}}$ (hot-spot control, larval-habitat management, entomological cordon).
    Recipient: continue to lower R_0 (IRS/LLIN, $\mu_m \nearrow$, rapid case management).

- $\boldsymbol{R_0} > 1$ (higher a, lower μ_m, high vector density, etc.)

Trajectory: self-sustained establishment (endemic equilibrium); importation accelerates take-off.
Action: intensive program to bring $R_0 < 1$ (high LLIN coverage, mass IRS, habitat elimination, accelerated care).

- **Implications for the low-transmission zone (action plan)**
  - Lower local R_0

    $a \searrow$: LLIN, IRS, repellents, physical barriers $\rightarrow$ fewer bites.
    $\beta \searrow$, $\beta_m \searrow$: rapid case management (shorter and weaker infectiousness), targeted treatment.
    $\mu_m \nearrow$: vector control (effective IRS, larviciding, water management) $\rightarrow$ shorter mosquito lifespan.
    Urban design: fewer breeding sites (stagnant water), improved housing $\rightarrow$ indirect effects on a and N_m

  - Reduce the infected inflow (act at the source)

    Hot spot: targeted IRS/LLIN, intensive habitat management, mass treatment if appropriate.
    Entomological frontiers: intervene along wind/mobility corridors (barriers, traps, coordinated operations).

- Align with seasonality

  Strengthen measures in the pre-season (rains, high-altitude winds); time logistics for periods when $\boldsymbol{I_{mh}}$ is likely to rise.

- Sensitive surveillance (key indicators)

  Entomological: parity, sporozoite rates, biting density (EIR), larval indices.
  Clinical: weekly incidence, care-seeking delays.
  Alerts: adaptive triggers when $R_0 \rightarrow 1^-$ or when winds/breeding conditions signal a surge in importation.

4.3 *Endemic Equilibrium*

At endemic equilibrium, the derivatives with respect to time are zero, i.e.,

$$\frac{dS_h}{dt} = \frac{dI_h}{dt} = \frac{dR_h}{dt} = \frac{dS_m}{dt} = \frac{dI_m}{dt} = 0$$

$$\begin{cases} \frac{dS_h}{dt} = 0 \\ \frac{dI_h}{dt} = 0 \\ \frac{dR_h}{dt} = 0 \\ \frac{dS_m}{dt} = 0 \\ \frac{dI_m}{dt} = 0 \end{cases} => \begin{cases} \wedge_h N_h - \left(\frac{a\beta I_m}{N_h} + \mu_h\right)S_h + \nu R_h = 0 \\ \frac{a\beta I_m}{N_h} S_h - (\mu_h + \gamma + b)I_h = 0 \\ \gamma I_h - (\mu_h + \nu)R_h = 0 \\ \mu N_m - \left(\frac{a\beta_m I_h}{N_h} + \mu_m\right)S_m + m(Im_h - Sm_l) = 0 \\ \frac{a\beta_m I_h}{N_h} S_m - \mu_m I_m = 0 \end{cases}$$

The solutions of these equations give the equilibrium values $S_h^*, I_h^*, R_h^*, S_m^*, I_m^*$.

To find the Jacobian matrix, we don't need the exact expressions but the equilibrium values.

$S_h^*, I_h^*, R_h^*, S_m^*, I_m^*$ are the stable solutions to these equations.

The Jacobian matrix J at endemic equilibrium is formed by the partial derivatives of the model equations with respect to the variables $S_h, I_h, R_h, S_m I_m$.

Combining all these partial derivatives, the Jacobian matrix J at endemic equilibrium is

$$J(E^*) = \begin{bmatrix} -a\beta\frac{I_m^*}{N_h} - \mu_h & 0 & \nu & 0 & -a\beta\frac{S_h^*}{N_h} \\ a\beta\frac{I_m^*}{N_h} & -(\gamma + \mu_h + b) & 0 & 0 & a\beta\frac{S_h^*}{N_h} \\ 0 & \gamma & -(\mu_h + \nu) & 0 & 0 \\ 0 & -a\beta_m S_m \frac{1}{N_h} & 0 & -a\beta_m\frac{I_h^*}{N_h} - (\mu_m + m) & m \\ 0 & a\beta_m S_m \frac{1}{N_h} & 0 & a\beta_m\frac{I_h^*}{N_h} & -\mu_m \end{bmatrix}$$

- **Analysis of the stability of the endemic balance**

To simplify the analysis, let's note:

$\lambda_1 = -a\beta\frac{I_m^*}{N_h} - \mu_h$, $\lambda_2 = -(\gamma + \mu_h + b)$, $\lambda_3 = -(\mu_h + \nu)$

These three terms are the eigenvalues corresponding to the first three rows and columns of the Jacobian matrix, as these rows and columns are decoupled from the others.

This leaves a 2×2 submatrix

$$J' = \begin{bmatrix} -a\beta_m\frac{I_h^*}{N_h} - (\mu_m + m) & m \\ a\beta_m\frac{S_m^*}{N_h} & -\mu_m \end{bmatrix}$$

The eigenvalues of this submatrix can be found by solving the characteristic equation:

$$\det\left(J' - \lambda I\right) = 0$$

$$Det\begin{bmatrix} -a\beta_m\frac{I_h^*}{N_h} - (\mu_m + m) - \lambda & m \\ a\beta_m\frac{S_m^*}{N_h} & -\mu_m - \lambda \end{bmatrix} = 0$$

$$\left(-a\beta_m\frac{I_h^*}{N_h} - (\mu_m + m) - \lambda\right)(-\mu_m - \lambda) - ma\beta_m\frac{S_m^*}{N_h} = 0$$

Let's develop

$$\lambda^2 + \lambda\left(-a\beta_m\frac{I_h^*}{N_h} + (\mu_m + m) + \mu_m\right) + (a\beta_m\frac{I_h^*}{N_h} + (\mu_m + m))\mu_m - ma\beta_m\frac{S_m^*}{N_h} = 0$$

$$(\lambda + \mu_m)\left(\lambda + \mu_m + m + a\beta_m\frac{I_h^*}{N_h}\right) - ma\beta_m\frac{S_m^*}{N_h} = 0$$

be the following quadratic

$$\boxed{\lambda^2 + p\lambda + q = 0}$$

With

$$p = a\beta_m\frac{I_h^*}{N_h} + 2\mu_m + m > 0,$$

$$q = \mu_m\left(\mu_m + m + a\beta_m\frac{I_h^*}{N_h}\right) - ma\beta_m\frac{S_m^*}{N_h}$$

– ***q* positivity at the endemic point**

At the endemic point (closed system, no external input), the stationary equation of I_m gives

$$\dot{I}_m = 0 \Rightarrow a\beta_m\frac{I_h^* S_m^*}{N_h} = \mu_m I_m^* \Rightarrow a\beta_m\frac{S_m^*}{N_h}I_h^* = \mu_m I_m^*$$

Then

$$a\beta_m\frac{S_m^*}{N_h} = \mu_m\frac{I_m^*}{I_h^*}$$

Thus

$$q = \mu_m\left(\mu_m + m + a\beta_m\frac{I_h^*}{N_h}\right) - ma\beta_m\frac{S_m^*}{N_h} = \mu_m(\mu_m + m + a\beta_m\frac{I_h^*}{N_h} - m\frac{I_m^*}{I_h^*})$$

Now the stationary equation of S_m (with the term $+mI_m$ that you included in J') is written as follows:

$$0 = \mu N_m - \underbrace{a\beta_m \frac{S_m^*}{N_h} I_h^*}_{\mu_m I_m^*} - (\mu_m + m)S_m^* + mI_m^*$$

Hence

$$\begin{aligned}\left(\mu_m + m + a\beta_m \frac{I_h^*}{N_h}\right) S_m^* &= \mu N_m + mI_m^* - a\beta_m \frac{S_m^*}{N_h} I_h^* \\ &= \mu N_m + mI_m^* - \mu_m I_m^* = \mu N_m - (\mu_m - m)I_m^*\end{aligned}$$

The left-hand side is positive (because $S_m^* > 0$), so the right-hand side is also, which imposes in particular

$$\mu_m + m + a\beta_m \frac{I_h^*}{N_h} > m \frac{I_m^*}{I_h^*}$$

Therefore, the above bracket is strictly positive and, as $\mu_m > 0$,

$$\boxed{q > 0}$$

– **Eigenvalues of** 2×2**and stability**

The two eigenvalues of J are

$$\lambda_{4,5} = \frac{-p \pm \sqrt{p^2 - 4q}}{2}$$

Since $\boldsymbol{p} > 0$ and $\boldsymbol{q} > 0$, we always have $\Re(\lambda_{4,5}) = -\frac{p}{2} < 0$ (negative real roots if $\boldsymbol{p}^2 \geq 4\boldsymbol{q}$, or complex conjugates with real part $-\frac{p}{2}$ otherwise).

Overall Conclusion

The three "human" eigenvalues are strictly negative.

The two eigenvalues of the mosquito block have negative real parts (since $p > 0$, $q > 0$).

$$\boxed{\text{The endemic equilibrium is locally asymptotically stable}}$$

5 Simulations

In this section, a simulation is carried out with supporting graphs, presenting the population dynamics of humans and mosquitoes in two distinct areas (recall: one area of high transmission and one area of low transmission of malaria.

5.1 Simulation of Human and Mosquito Population Dynamics in High- and Low-Transmission Zones

- **Human population dynamics:**
  - **Infected Human (High Transmission)** (solid orange line): The number of infected humans in the high-transmission area decreases rapidly in the early days, then stabilizes at a near-zero level. This indicates that the disease is controlled in this area after an initial infection peak (Fig. 2).
  - **Infected Human (Low Transmission)** (dashed orange line): Similarly, in the low-transmission area, the number of infected humans also drops quickly before stabilizing at a very low level.

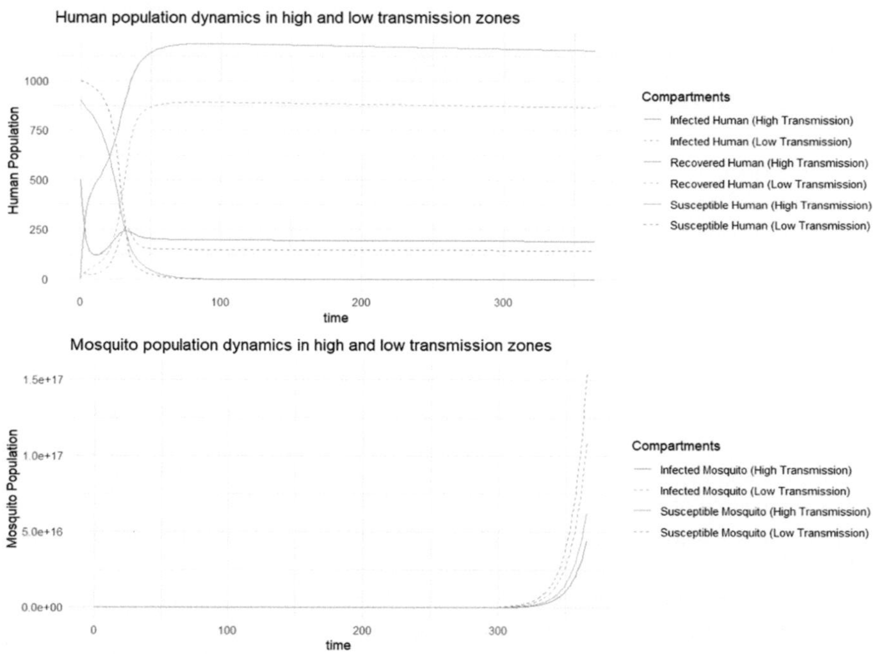

Fig. 2 Simulation of human and mosquito population dynamics in high- and low-transmission zones

- **Recovered Human (High Transmission)** (solid green line): The number of recovered humans in the high-transmission area increases rapidly in the first days and reaches a constant value, showing that a large part of the infected population has recovered.
- **Recovered Human (Low Transmission)** (dashed green line): In the low-transmission area, a similar behavior is observed, with a growing number of recovered humans reaching a plateau.
- **Susceptible Human (High Transmission)** (solid blue line): The population of susceptible humans in the high-transmission area decreases slightly at first, then increases again, likely due to the control of the infection and an increase in recoveries.
- **Susceptible Human (Low Transmission)** (dashed blue line): In the low-transmission area, susceptible humans follow a similar path with an initial drop before gradually increasing.

• **Mosquito population dynamics:**

- The mosquito dynamics seem to exhibit a particular behavior toward the end of the period (around day 300), where the mosquito population increases exponentially in both areas, for both infected and susceptible mosquitoes.
- **Infected Mosquito (High Transmission)** (solid red line): The number of infected mosquitoes in the high-transmission area remains stable until around day 300, then increases rapidly, indicating an accelerated spread of the disease in the mosquito population.
- **Infected Mosquito (Low Transmission)** (dashed red line): A similar trend is observed in the low-transmission area, with an exponential increase in infected mosquitoes.
- **Susceptible Mosquito (High Transmission)** (solid purple line) and **Susceptible Mosquito (Low Transmission)** (dashed purple line): Susceptible mosquitoes in both areas show a similar drastic increase after a stable period.

5.2 *Simulation of Mosquito Movements*

Although the impact on humans is moderate, the movement of infected mosquitoes can lead to a dramatic spread of the disease, which in the long term could pose an increased risk to the human population in the low-transmission area. We demonstrate through the simulation that effective control strategies (LLINs, larval control or larvicides, etc.) can reduce the population of infected mosquitoes and limit their ability to move between transmission zones (Fig. 3).

The graph shows the results of the simulation with a significant movement of mosquitoes from the high-transmission area to the low-transmission area, and its impacts on the populations of infected humans and mosquitoes in the low-transmission area.

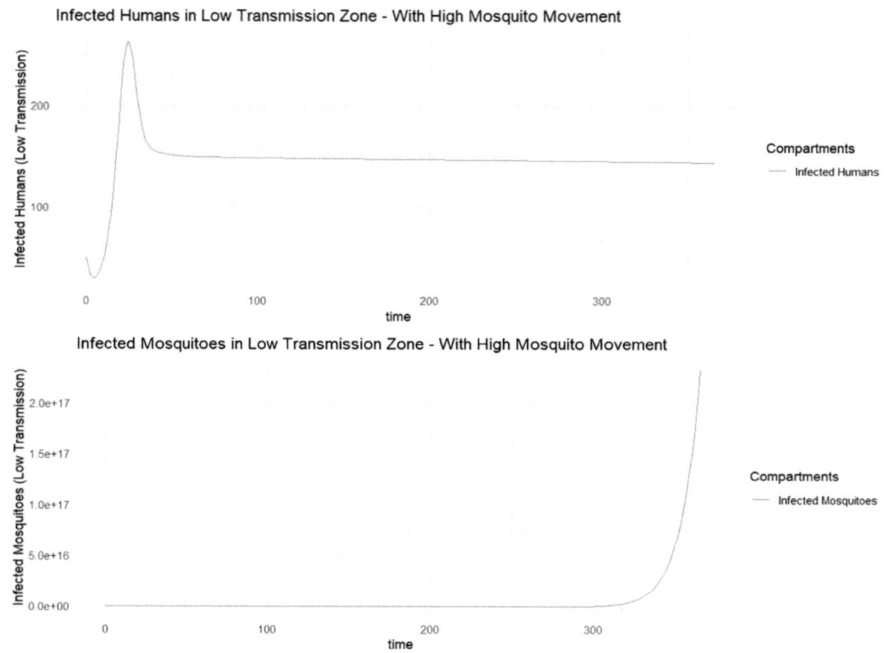

Fig. 3 Simulation of mosquito movements

In Infected Humans in Low-Transmission Zone—With High Mosquito Movement

The number of infected humans in the low-transmission area shows an initial peak, reaching around 250 infected individuals, followed by a rapid drop and stabilization around 150 infected individuals. Despite the increased movement of infected mosquitoes into this area, transmission among humans remains relatively controlled after the initial period. This suggests that, although the movement of mosquitoes has an impact, the measures in place prevent a continuous outbreak of human infections in this area.

In Infected Mosquitoes in Low-Transmission Zone—With High Mosquito Movement

The number of infected mosquitoes in the low-transmission area remains relatively stable until around day 300, at which point a very marked exponential growth is observed. By the end of the simulation period (day 365), the number of infected mosquitoes reaches extremely high levels.

5.3 Simulation of Mosquito Movement Scenarios

Here, scenario 2 shows a drastic reduction in the number of mosquito bites (Fig. 4).

The graph presents two scenarios for the low-transmission area:

- **Scenario without mosquito movement**: An initial infection peak is observed among humans in the low-transmission area, reaching around 250 infected individuals. This peak is followed by a rapid decline before the infected population stabilizes around 150 individuals. The infection remains stable after this drop, indicating that without the movement of mosquitoes from the high-transmission area, the transmission dynamics reach a relatively stable equilibrium in the low-transmission area.
- **Scenario with mosquito movement**: The scenario with mosquito movement shows similar behavior to the one without movement. There is an initial infection peak, followed by stabilization at a similar level (around 150 infected individuals). However, it seems that the movement of mosquitoes did not have a visible effect on the infection curve in the low-transmission area.

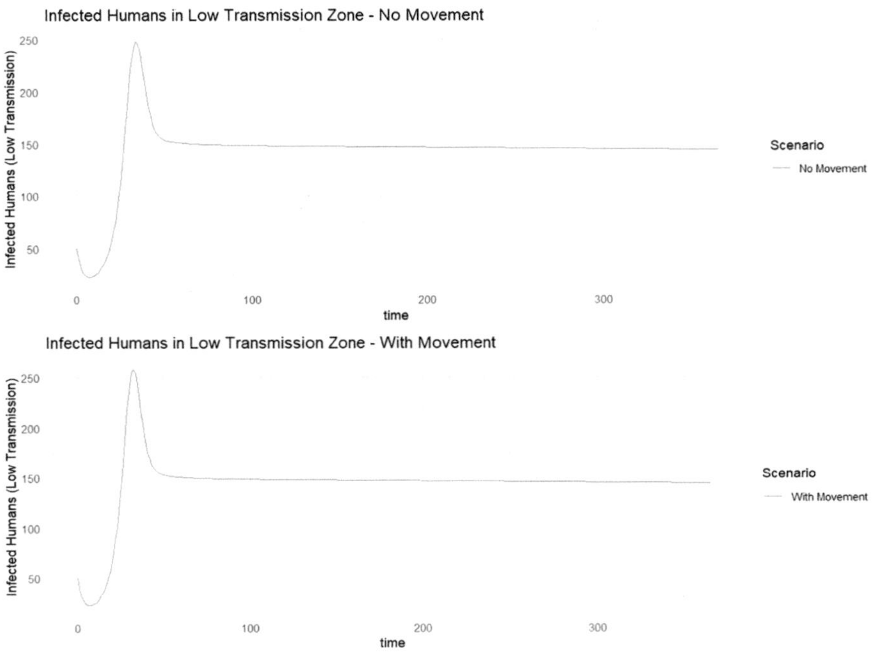

Fig. 4 Simulation of travel scenarios

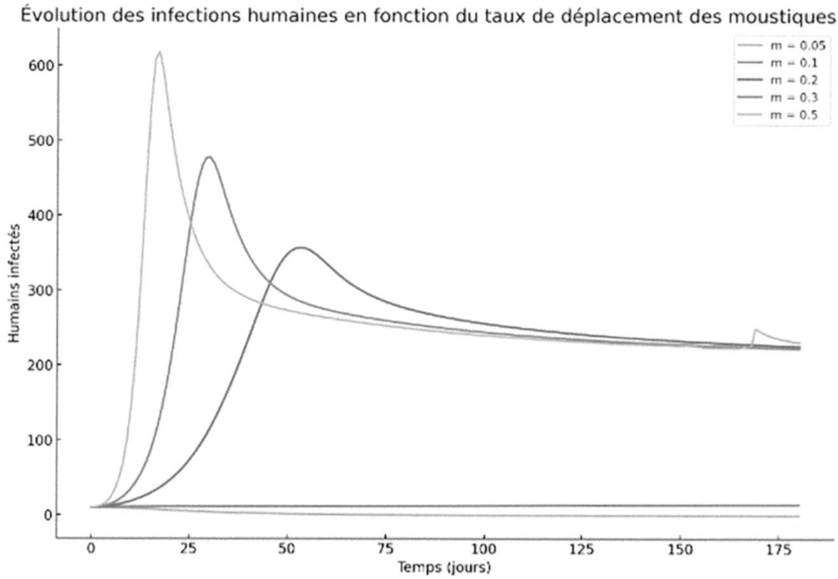

Fig. 5 Evolution of human infections as a function of the mosquito movement rate (m)

5.4 *Evolution of Human Infections as a Function of the Mosquito Movement Rate (M)*

Here, we perform simulations based on the parameter m, which represents the rate of mosquito movement between the two different areas, to analyze its impact on the evolution of malaria transmission. By taking increasing values of m, we study how mosquito mobility influences the dynamics of infections in both mosquitoes and humans. We visualize the results for each value of mmm, which allows us to identify trends and assess the effect of vector migration on the spread of the disease (Fig. 5).

The analysis of the simulation results as a function of the mosquito movement rate mm shows that when mm is low ($m = 0.05$), disease transmission is relatively stable and decreases slowly, since mosquitoes remain in one area and spread the infection locally. As mm increases, the spread becomes faster, reaching a higher peak before declining, which indicates that mosquito mobility intensifies transmission between different areas and accelerates the diffusion of infection.

For $m = 0.5$, the initial spread is very strong, but the decline is also faster due to the dispersion of the infectious load, which prevents any single area from maintaining sustained prevalence, thereby highlighting the impact of mosquito migration on malaria transmission dynamics.

6 Discussion

Our work develops an extended SIR model that makes mosquito mobility between high-transmission (hot spot) and low-transmission areas explicit, in order to assess how these flows modulate the disease-free and endemic equilibria. The framework highlights two mechanisms:

(i) a local threshold governed by R_0, such that $R_0 < 1$ guarantees the extinction of small introductions, whereas $R_0 > 1$ 1 allows the establishment of an endemic equilibrium E^* which, in the closed system (no external input), is locally asymptotically stable;

(ii) an intrinsically ambivalent mobility effect mm: in the closed system, mm reduces the residence time of susceptible vectors in the recipient area ($S_m^* = \frac{\mu}{\mu_m+m} N_m$) and therefore decreases R_0, but in the presence of an exogenous inflow of infected mosquitoes from the hot spot (mI_{mh}), the same mobility acts as a forcing capable of maintaining a low-level endemic state even when $R_0 < 1$, especially near the threshold. This reading reconciles earlier empirical and theoretical results: dispersal can sustain persistence (Cosner et al., 2009; Smith et al., 2004) while exhibiting non-linear effects whereby moderate migration maximizes apparent prevalence (Reiner et al., 2013; Macdonald, 1957), whereas too little or too much mobility can, depending on context, localize transmission or dilute it.

In contrast to classical SIR models, our approach links mosquito migration directly to the spatial distribution of infection, showing why localized interventions (LLIN/IRS, larval-habitat management) may be insufficient if entomological connectivity is not addressed—consistent with observations in settings of insecticide resistance (Hemingway et al., 2016) and with the importance of environmental and climatic factors that modulate mobility (Parham et al., 2012). In sum, our results frame our research question as follows: to quantify and disentangle the structural effect (via R_0,) and the forcing effect (via mI_{mh}) of vector movements in order to explain persistence or resurgence of malaria in low-transmission areas and to guide strategies that combine lowering local R_0, with controlling inflows from hot spots.

Limits of the Study

- Simplified compartmental model (no latent stages or age structure).
- Parameters assumed constant: no seasonality, climate, behavioral change, or insecticide resistance.
- Vector mobility aggregated into a single parameter (no distinction among local movement, windborne migration, or landscape effects).
- Human mobility not modeled (outbreak synchronization and reseeding ignored).
- Strong demographic assumptions (constant human population, linear mosquito demography, no density dependence).
- Analysis limited to local stability; no study of global stability, bifurcations, or sustained cycles.

- Calibration and uncertainty: incomplete identifiability and no propagation of stochasticity or measurement error.

7 Conclusion

This study provides an analytical and numerical framework for understanding how mosquito mobility between high- and low-transmission areas shapes malaria dynamics. By extending the SIR model to include a mobile vector component, we establish a local threshold governed by R_0: $R_0 < 1$ implies extinction of introductions, whereas $R_0 > 1$ leads to the existence and local stability of an endemic equilibrium. We highlight the dual role of mobility mm: structurally protective by reducing the residence of susceptible vectors ($S_m^* = \frac{\mu}{\mu_m+m}N_m$) and thus lowering R_0, yet potentially sustaining transmission through an importation forcing term (mI_{mh}) that can maintain low-level endemicity near the threshold. Simulations corroborate these results and show that sensitivity to imported fluxes increases as $R_0 \to 1^-$. Operationally, our findings argue for coupled strategies: lowering local R_0 (reducing mosquito–human contact, shortening vector survival, rapid case management) and acting at "entomological frontiers" to mitigate inflows from hot spots, with seasonally tuned interventions and responsive surveillance. This framework thus offers a robust conceptual tool to anticipate persistence or resurgence in low-transmission settings and to guide more targeted and coordinated control policies.

Future Directions

- Extend to a multi-zone metapopulation framework (networks, gravity/radiation), with explicit seasonality and high-altitude winds.
- Introduce stochastic processes for low-incidence phases, and perform data assimilation (entomological + clinical) to estimate $m(t)$, $a(t)$, and $\mu_m(t)$.
- Incorporate human mobility and test coordinated "source–recipient" strategies.
- Study the identifiability of key parameters and quantify forecast uncertainty.

References

1. Ross, R.: The Prevention of Malaria. John Murray (1911)
2. Macdonald, G.: The Epidemiology and Control of Malaria. Oxford University Press (1957)
3. Diekmann, O., Heesterbeek, J.A.P.: Mathematical Epidemiology of Infectious Diseases: Model Building. Wiley, Analysis and Interpretation (2000)
4. Chitnis, N., Hyman, J.M., Cushing, J.M.: Determining important parameters in the spread of malaria through the sensitivity analysis of a mathematical model. Bull. Math. Biol. **70**(5), 1272–1296 (2008). https://doi.org/10.1007/s11538-008-9299-0
5. Smith, D.L., et al.: Ross, macdonald, and a theory for the dynamics and control of mosquito-transmitted pathogens. PLoS Pathog. **5**(4), e1000051 (2004). https://doi.org/10.1371/journal.ppat.1002588

6. Ferguson, N.M., et al.: Strategies for containing an emerging influenza pandemic in Southeast Asia. Nature **437**(7056), 209–214 (2005). https://doi.org/10.1038/nature04017
7. Juarez, J.G., et al.: Dispersal of female and male Aedes aegypti from emergence in a multi-block field site. Sci. Rep. **10**, 6803 (2020)
8. Amaro, F.I.F. et al.: Mark–Release–Recapture Trial with Aedes albopictus (Diptera, Culicidae) Irradiated Males: Population Parameters and Climatic Factors (2024)
9. Huestis, D.L., et al.: Windborne long-distance migration of malaria mosquitoes in the Sahel. Nature **574**, 404–408 (2019)
10. Sinka, M.E., et al.: A new malaria vector in Africa: predicting the expansion range of Anopheles stephensi and identifying the urban populations at risk. PNAS **117**(40), 24900–24908 (2020)
11. Emiru, T. et al.: Evidence for a role of Anopheles stephensi in the spread of urban malaria in Africa. Nature Medicine (2023)
12. Taylor, R. et al.: Invasive Anopheles stephensi in Africa: insights from Asia. Trends in Parasitology (2024)
13. Lee, S.A., et al.: Spatial connectivity in mosquito-borne disease models: a systematic review. J. R. Soc. Interface. **18**, 20210096 (2021)
14. Citron, D.T., et al.: Comparing metapopulation dynamics of infectious disease models. PNAS **118**(51), e2007488118 (2021)
15. Wang, J., et al.: Analysis of a diffusive malaria model with drug-sensitive and resistant strains. J. Biol. Dyn. **18**(1), 1–21 (2024)

Numerical Simulation by Finite Element Method of the Deformation of an Embedded Plate

Mamadou Diop, Malicki Zorom, Babacar Leye, Malick Pane, Ibrahima Mbaye, and Adamah Messan

Abstract This work aims to study the influence of soil properties on the deformation of a foundation plate resting on an elastic soil. A model based on the Vlasov-foundation theory is proposed. First, we have shown the existence and uniqueness of the solution of the problem by using the Lax–Milgram theorem. Then, numerical simulations by finite elements are carried out with the FreeFem++ software to determine the deformations of the plate depending on the different values of the soil and plate parameters. This study has also shown that the parameters of the sub-grade are more influential on the deformation of the plate than those of the plate itself.

Keywords Lax–Milgram · Finite element · Foundation · FreeFem++ · Soil-structure interaction · Plate · Model of Vlasov

1 Introduction

The study of plate type structures resting on elastic foundations is a very important research subject not only in the field of mechanics but also in the field of civil engineering. Plates on elastic foundations represent complex soil-structure interaction problems. The development of equations for such problems and the associated solutions have become a challenge for engineers and mathematicians [1–7]. In fact all the loads of the structure must be transferred to the ground, and the whole ground and structure must act together to resist and support the loads. The integral nature

M. Diop (✉) · A. Messan
Laboratoire Eco-Matériaux et Habitats Durables (LEMHaD), Institut International d'Ingénierie de l'Eau, et de l'Environnement (2iE), Rue de La Science, Ouagadougou, Burkina Faso
e-mail: mamadou.diop@2ie-edu.org

M. Zorom · B. Leye
Laboratoire Eaux Hydro-Systèmes et Agriculture (LEHSA), Institut International d'Ingénierie de l'Eau, et de l'Environnement (2iE), Rue de La Science, Ouagadougou, Burkina Faso

M. Pane · I. Mbaye
Département de Mathématiques, UFR SET, Université Iba Der Thiam (UIDT), Thiès, Sénégal

© The Author(s), under exclusive license to Springer Nature Switzerland AG 2026

C. T. Gueye et al. (eds.), *Mathematics, Computer Science, Cybersecurity and Artificial Intelligence*, Springer Proceedings in Mathematics & Statistics 536,
https://doi.org/10.1007/978-3-032-17666-0_8

of the foundation and soil action is made further complicated by the complexity of the soil environment itself. Various analytical methods have been employed to find effective factors regarding foundation-plate behavior for practical applications in civil/structural engineering. The analysis of several cases on the vibration of rectangular plates is considered in [8–13, 16]. The plate theory also plays an important role in the simulation of numerical models and the structure design of the plate-foundation system. The development of the Vlassov mathematical model is introduced for this type of very complex soil-structure interaction problem in order to provide designers with realistic values in the design. The mathematical model takes into account two parameters, the vertical soil reaction modulus k and the horizontal soil reaction modulus 2T which depend on the geometric and mechanical characteristics of the system. From a practical perspective, numerical methods, in particular the finite element method, are essential for the dimensioning and the verification of this type of complex structures. Some models such as Winkler and Pasternak foundations have been introduced to consider interactions between the plate and its foundation. Other improved models have been proposed to develop basic models, particularly by considering nonlinear dynamic aspects of plate foundation systems [17–19]. However, few studies have rigorously combined a mathematical analysis of existence and uniqueness with numerical simulations based on the finite element method. In this work, a model based on the Vlasov-foundation theory is proposed. Through this study, we have first shown the existence and the uniqueness of the solution of the mathematical model considered by means of the Lax–Milgram theorem [14]. Then the software FreeFem++ [15] is used to determine the deformations of the plate as a function of the different values of the soil and plate parameters.

2 Presentation of the Model

In the analysis of foundations, the soil is considered as a set of springs of elastic modulus k infinitely close to each other and connected by an elastic membrane of horizontal reaction modulus equal to 2T. The following figure shows the representation of a loaded plate.

The theory of plates on an elastic foundation taking into account the soil-structure interaction leads to the following Vlasov model:

Find w: $\Omega \subset \mathbb{R}^2 \to \mathbb{R}$ the deflection of the plate such that

$$\begin{cases} D\nabla^4 w(x, y) - 2T\nabla^2 w(x, y) + kw(x, y) = q(x, y), \forall (x, y) \in \Omega \\ w = \frac{\partial w}{\partial \eta} = 0 \qquad on\ \partial\Omega \end{cases} \tag{1}$$

where $w(x, y)$ is the deflection of the plate, $q(x, y)$ is the loadings applied on surface of the plate, D is the bending stiffness of the plate and is given by

$$D = \frac{E_p e^3}{12(1 - \nu_p^2)}.$$

E_p is the modulus of elasticity of the material constituting the plate, e is the thickness of the plate, ν_p is the Poisson's ratio of the plate, k is the soil reaction modulus, T is the shear parameter, $q \in \mathbb{L}^2(\Omega)$ is the uniformly applied load, η is the outside unit normal vector of Ω, and H is the depth of the soil layer.

Many researchers have worked to develop a technique for determining the modulus of reaction k [23]. In this article, we consider

$$k = \frac{E_s(1 - \nu_s)}{(1 + \nu_s)(1 - 2\nu_s)} \int_0^H \left(\frac{d\phi}{dz}\right)^2 dz \quad (2)$$

$$2T = \frac{E_s}{2(1 + \nu_s)} \int_0^H \phi^2 dz \quad (3)$$

with,

E_s is the modulus of elasticity of the subgrade, ν_s is the Poisson's ratio of the foundation soil, and ϕ is the soil deformation mode shown in Fig. 1.

Selvadurai (1979) suggests two expressions of the function ϕ with its boundary conditions $\phi(0) = 1$ and $\phi(H) = 0$ defined by:

$$\phi(z) = (1 - z/H) \quad (4)$$

$$\phi(z) = \frac{\sinh\gamma(1 - z/H)}{\sinh\gamma} \quad (5)$$

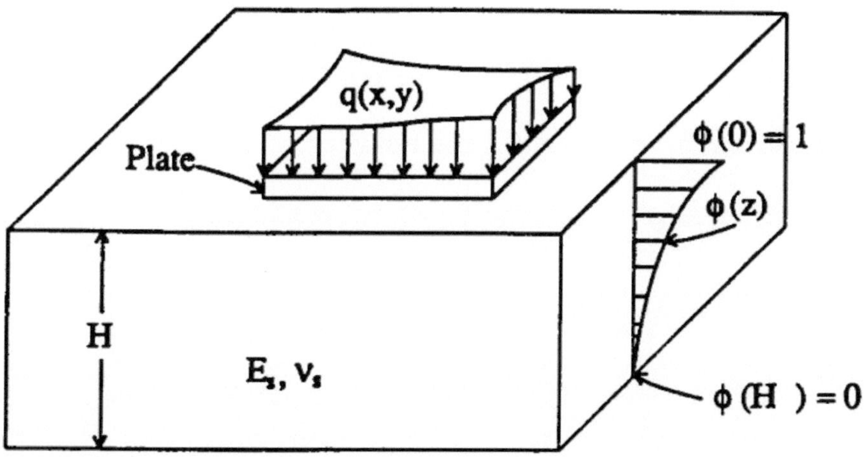

Fig. 1 Representation of a loaded plate resting on an elastic foundation

2.1 Existence and Uniqueness of the Problem Solution Eq. (1)

Proposition 1: According to Lax–Milgram theorem the problem Eq. (1) admits a unique solution.

$w \in V = \mathbb{H}_0^2(\Omega) = \left\{ v \in \mathbb{H}^2(\Omega) : v = \frac{\partial v}{\partial \eta} = 0 \right\}$ verifying the following variational formulation:

$$\text{Find } w \in V \text{ such that } a(w, v) = b(v) \forall v \in V \tag{6}$$

With $a : V \times V \to \mathbb{R}$ defined by:

$$a(w, v) = D \int\limits_{\Omega} \Delta w \Delta v dxdy + 2T \int\limits_{\Omega} \nabla w \nabla v dxdy \; + k \int\limits_{\Omega} wvdxdy$$

and $b : V \to \mathbb{R}$ defined by:

$$b(v) = \int_{\Omega} qvdxdy$$

We provide V the reduced norm of $\mathbb{H}_0^2(\Omega)$ defined by:

$$b(v) = \int\limits_{\Omega} qv \, dx \, dy$$

The space V provided with the norm $\|.\|_V$ is a Hilbert space.

Proof Since the map a is bilinear and symmetric by definition, likewise b is linear by definition, it suffices to show that a is continuous and coercive and that b is also continuous. Indeed, we will verify the assumptions of Lax–Milgram theorem [1]:

Continuous of a: We have for all $(w, v) \in V \times V$:

$$|a(w, v)| = \left| D \int_{\Omega} \Delta w \Delta v dxdy + 2T \int_{\Omega} \nabla w \nabla v dxdy + k \int_{\Omega} wvdxdy \right|$$

$$\leq D \left| \int_{\Omega} \Delta w \Delta v dxdy \right| + 2T \left| \int_{\Omega} \nabla w \nabla v dxdy \right| + k \left| \int_{\Omega} wvdxdy \right|$$

$$\leq D\|w\|_V \|v\|_V + 2T \|\nabla w\|_{\mathbb{L}^2(\Omega)} \|\nabla v\|_{\mathbb{L}^2(\Omega)} + k\|w\|_{\mathbb{L}^2(\Omega)} \|v\|_{\mathbb{L}^2(\Omega)}$$

$$\leq D\|w\|_V \|v\|_V + 2TC^2 \|w\|_V \|v\|_V + kC^2 \|w\|_V \|v\|_V$$

$$|a(w,v)| \leq \left(D + 2TC^2 + kC^2\right)\|w\|_V \|v\|_V.$$

Therefore a is continuous on V where C is the constant of Poincaré-Friedrichs.

Coercivity of a: we have for all $w \in V$:

$$a(w,w) = D\int_{\Omega}(\Delta w)^2 dxdy + 2T\int_{\Omega}(\nabla w)^2 dxdy + k\int_{\Omega} w^2 dxdy$$

where $a(w,w) \geq D\|w\|_V^2$

therefore a is coercive. We will prove that b is continuous.

Continuous of b: Indeed, we have for all $v \in V$:

$$|b(v)| = \left|\int_{\Omega} qvdxdy\right|.$$

Cauchy Schwarz inequality leads to:

$$|b(v)| \leq \|q\|_{\mathbb{L}^2(\Omega)}\|v\|_{\mathbb{L}^2(\Omega)}$$

by using Poincaré inequality, we obtain:

$$|b(v)| \leq C\|q\|_{\mathbb{L}^2(\Omega)}\|v\|_V$$

therefore b is continuous.

The Assumptions of the Lax–Milgram theorem are verified therefore the variational problem Eq. (7) admits a unique solution therefore Eq. (1) also admits a unique solution.

3 Approximation by Finite Element Method and Numerical Simulation

In this section, we first present the finite element method based on the internal approximation construction of $\mathbb{H}_0^2(\Omega)$ by N_T triangles, then we determine the numerical solutions of the problem with the FreeFem++ software.

3.1 Approximation by Finite Element Method

We assume that the domain Ω has a piecewise polygonal border so that it is completely covered by N_T small triangles T_k, $k \in \{1, 2, \ldots, N_T\}$ of maximum size h. We denote by $\mathbb{P}^2$ the set of polynomials of degree less than or equal to 2 in the mean plane (x, y).

The problem Eq. (1) is written in the following variational form: find $w \in V$ such that

$$a(w, v) = b(v), \forall v \in V.$$

The discrete variational space is written in the form:

$$V_h = \left\{ v_h \in \widehat{V_h}, v_{h|\partial\Omega} = \frac{\partial v_h}{\partial \eta}_{|\partial\Omega} = 0 \right\}$$

where

$$\widehat{V_h} = \left\{ v_h \in C^0(\Omega), v_{h_{|T_k}} \in \mathbb{P}^2(\Omega), \forall k = 1,2,\ldots,N_T \right\}$$

The Lax–Milgram theorem ensures the existence and the uniqueness of w_h the solution of the following approximate variational problem [3]: find $w_h \in V_h$ such that

$$a(w_h, v_h) = l(v_h), \forall v_h \in V_h.$$

3.2 Presentation of the Results

The different values of the simulation parameters are taken [1, 2]. Indeed, the type of plate used in this work is straight pad in width $l = 10\,m$, length $L = 12\,m$ and thickness $e = 20\,cm$ in the respective directions x, y , and z subjected to an evenly distributed load $q = 23.94\,MPa$ for a depth $H = 10\,m$.

The modulus of elasticity of the plate and the Poisson's ratio are respectively: $E_p = 43\,GPa$ and $\nu_p = 0.45$, the modulus of elasticity of the soil and the Poisson's ratio of the soil are respectively: $E_s = 8\,MPa$ and $\nu_s = 0.25$. The parameters k and $2T$ are calculated with $\phi(z) = (1 - z/H)$. We represent the mesh of the domain Ω in Fig. 2.

Figures 3 and 4 represent the deformation of the plate in 2D and 3D respectively.

The following Figs. 5, 6, 7, and 8 show the influence of the various parameters on the deformation of the plate on an elastic foundation.

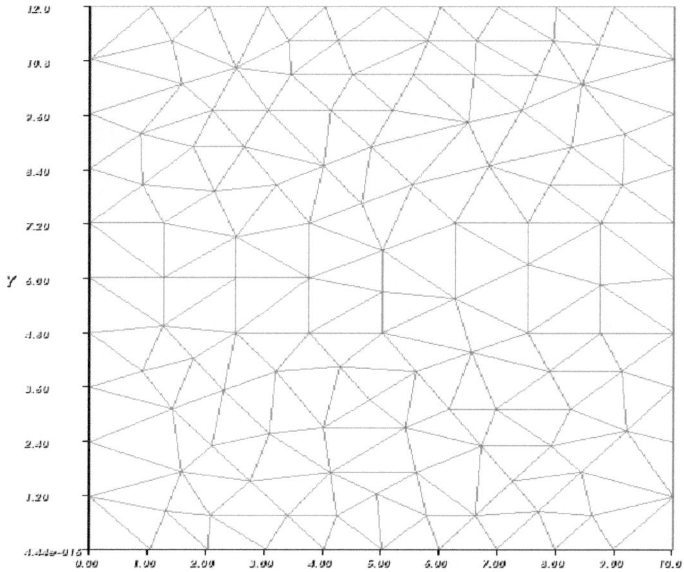

Fig. 2 Mesh of the domain

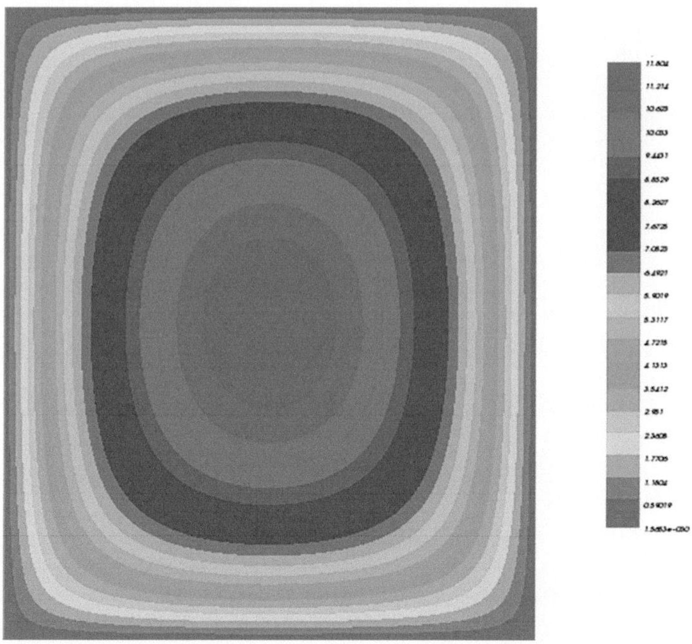

Fig. 3 2D plate deformation visualization with $E_p = 43\,GPa$, $v_p = 0.45$, $E_s = 8\,MPa$ and $v_s = 0.25$

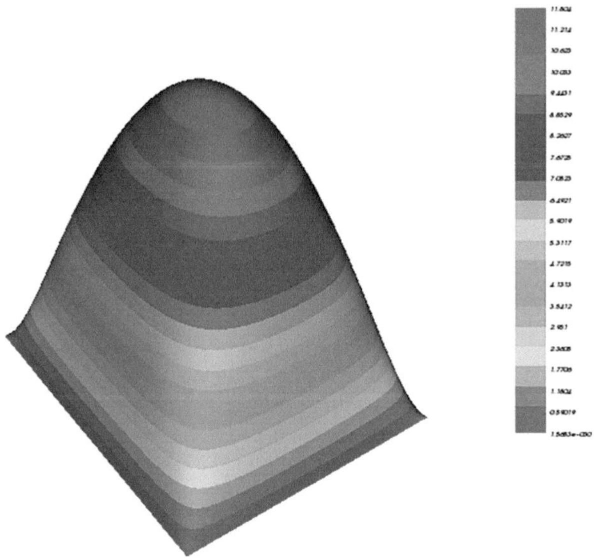

Fig. 4 3D plate deformation visualization *with* $E_p = 43\,GPa$, $v_p = 0.45$, $E_s = 8\,MPa$ and $v_s = 0.25$

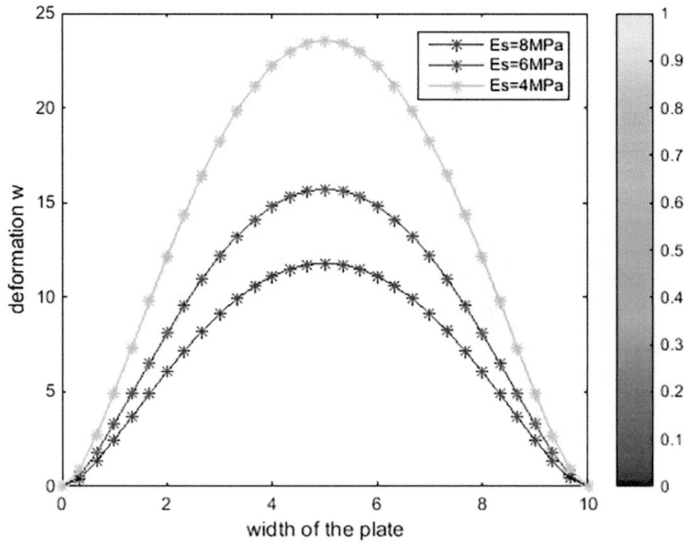

Fig. 5 Plate deformation for different values of E_s

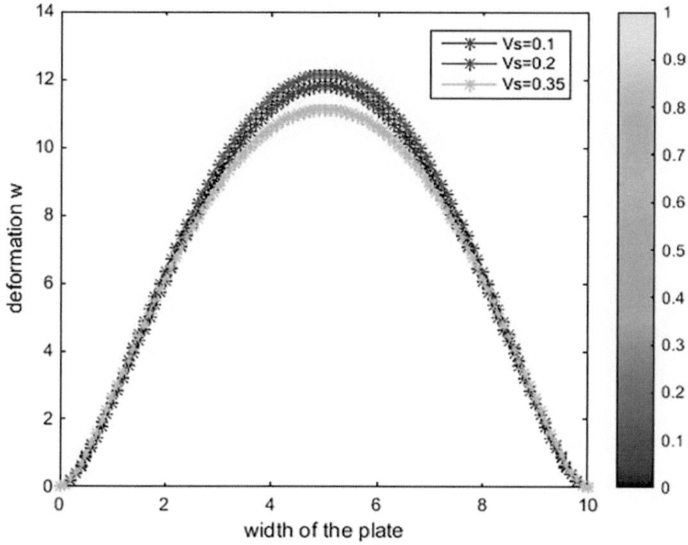

Fig. 6 Plate deformation for different values of v_s

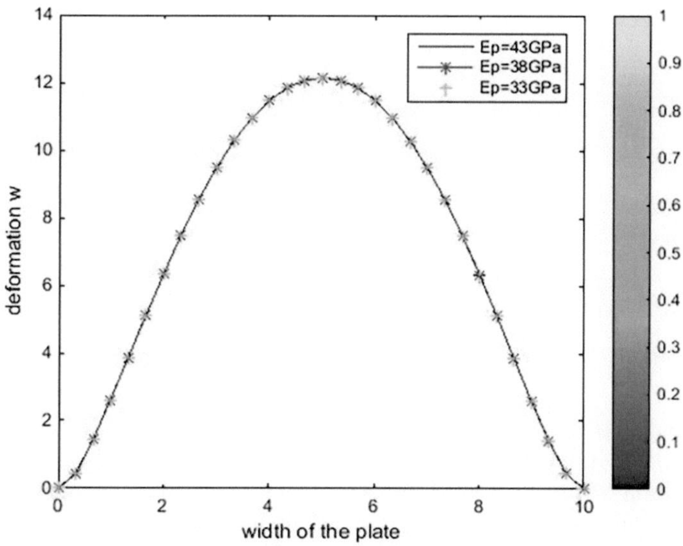

Fig. 7 Plate deformation for different values of E_p

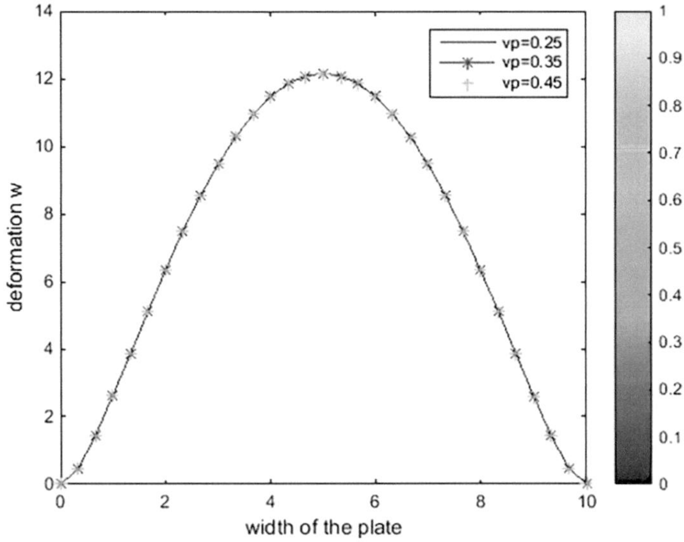

Fig. 8 Plate deformation for different values of v_p

Figure 5 reveals a significant sensitivity of the plate deformation to the elastic modulus E_s. The results highlight an inverse correlation between these two variables, indicating that the deformation increases as E_s decreases.

Figure 6 also shows that as v_s increases, the deformation becomes progressively smaller.

Figures 7 and 8 show that the variation of the parameters (E_p,v_p) has a negligible influence on the results.

3.3 Discussion of the Results

Figures 5 and 6 show by the finite element method that the soil parameters (E_s, v_s) have a significant influence on the plate deflection. However, Figs. 7 and 8 show that the deformation of the plate is almost insensitive to the plate parameters $\left(E_p, v_p\right)$. Following this analysis, it is concluded that the modulus of elasticity of the foundation soil has a very significant influence on the deformation of the plate, on the other hand the parameters of the plate have little influence. Thus, the study reveals that knowledge of the mechanical properties of the soil is a prerequisite for mastering the behavior of foundation structures. Moreover, the numerical results show that the deformations are not only a function of the loads but also a function of the mechanical properties of the subsoil and of the plate.

4 Conclusion

In this work, a mathematical model is considered to determine the deformations of an embedded plate. First, we have proved the existence and the uniqueness of the solution of the problem by the means of the theorem of Lax–Milgram and the results of analyzes, then numerical simulations by finite element are carried out with FreeFem++ software. Numerical results from finite elements calculations are verified by comparing with published results and show a good correlation. Thus, the numerical results found show a significant influence of the various parameters of the soil (E_s, ν_s) on the deformations of the plate. These results confirm the validity of the model and demonstrate its relevance for analyzing the behavior of plates on elastic foundations. The dynamic study of the model is a perspective for future work. Moreover, this work could be extended to simply supported, free, or plates with mixed boundary conditions.

References

1. Guellil, M. E, Tahar, A. M, Harrichane, K.: Modélisation du comportement des plaques reposant sur sol élastique par la méthode des éléments finis, In: 2eme Conférence Internationale sur le Calcul Numérique en Géotechnique; NUCGE, (2011)
2. Sall, O.A., Ba, M., Ndiaye, M., Fall, M., Berthaud, Y., Mbaye, I., Fall, M.: Finite element computation of the behavioral model of mat foundation. Int. J. Civ. Eng. Technol.; IJCIET, (2015)
3. Buczkowski, R., Torbacki, W.: Finite element modeling of thick plates on two-parameter elastic fondation. Int. J. Numer. Anal. Methods Geomech., (2001)
4. Mbaye, I., Diop, M., Sonko, A., Ba, M.: Analysis of the dynamic response of the soil-pile behavioral model under lateral load. Aust. J. Math. Anal. Appl., (2021)
5. Straughan, W.T.: Analysis of plate on elastic foundation. Thesis Universities de technologies du Texas, Ph-D (1990)
6. Mofid, M., Noroozi, M.: A plate on winkler foundation with variable coefficient. Sharif University of Technology, Tehran, Iran (2009)
7. Wu, J., Zhang, J., Li, X.: Bending analysis of stepped rectangular plates resting on an elastic half-space foundation. Buildings, (2023)
8. Lee, S., Lin, S.: Levy-type solution for the analysis of nonuniform plates. Comput. Struct, (1993)
9. Malekzadeh, P., Farid, M.: A DQ large deformation analysis of composite plates on nonlinear elastic foundation. Compos. Struct, (2007)
10. Laura, P., Gutierrez, R.: Vibrating non uniform plates on elastic foundation. J. Eng. Mech. ASCE, (1985)
11. Biswas, P.: Vibration of irregular-shaped orthotropic plates resting on elastic foundation under in-plane forces. J. Indian. Inst. Sci., (1978)
12. Tenenbaum, J., Eisenberger, M.: Analytic solution for buckling of rectangular isotropic plates with internal point supports. Thin Wall Struct, (2023)
13. Straughan, W.T.: B.S, and M.S. analysis of plate on elastic foundation. Ph-D, Universities de Technologies du Texas, (1990)
14. Allaire, G.: Analyse numérique et optimisation. Édition de l'école polytechnique, (2012)
15. Hertz, F.: New development FreeFem++. J. Numer. Math. Gruyter, (2012)

16. Szilard, R.: Theories and applications of plates analysis classical, numerical and engineering methods, Hoboken, USA (2004)
17. Younesian, D., Hosseinkhani, A., Askari, H., Esmailzadeh, E.: Elastic and viscoelastic foundations: a review on linear and nonlinear vibration modeling and applications. Nonlinear Dyn., (2019)
18. Kumar, S., Jana, P.: Accurate solution for free vibration behaviour of stepped FGM plates implementing the dynamic stiffness method. Structures, (2023)
19. Singh, S.J, Harsha, S.P.: Nonlinear dynamic analysis of sandwich S-FGM plate resting on pasternak foundation under thermal environment. Eur. J. Mech.-A/Solids, (2019)

IoTs and AI for Sustainable Wastewater Management in Senegal: A Status Report

Damien Diop, Idy Diop, Massa Ndong, Nouhou Diaby, and Alassane Bah

Abstract This article provides a comprehensive review of the challenges and opportunities associated with wastewater management in Senegal. By examining current practices, existing infrastructure, and environmental and health impacts, we identify key gaps and areas for improvement. The study highlights ongoing efforts to adopt more sustainable solutions, while highlighting the socio-economic and technological constraints hindering their implementation. Finally, the article discusses innovative approaches, including the integration of Information and Communication Technologies (ICT) and community-based management initiatives, as promising avenues for strengthening sustainable wastewater management in Senegal.

Keywords Wastewater · Climate change · Water quality analysis · Water distribution · Population growth · Rural areas · Water-related diseases

1 Introduction

Water scarcity has become a major global challenge, with many regions around the world experiencing increasing water shortages. Every year, 297,000 children under the age of five die from diarrheal diseases caused by unsafe water and inadequate sanitation practices [1, 2]. Furthermore, 80% of industrial and municipal wastewater is discharged into the environment without treatment [1, 3].

D. Diop (✉) · I. Diop · N. Diaby · A. Bah
Cheikh Anta Diop University, Dakar, Senegal
e-mail: diopdamien903@gmail.com

D. Diop · I. Diop · A. Bah
Science, Polytechnic Institute (ESP), Dakar, Senegal

M. Ndong
Cheikh Hamidou Kane University, Dakar, Senegal

N. Diaby
Fundamental Institute of Black Africa (IFAN), Dakar, Senegal

© The Author(s), under exclusive license to Springer Nature Switzerland AG 2026

C. T. Gueye et al. (eds.), *Mathematics, Computer Science, Cybersecurity and Artificial Intelligence*, Springer Proceedings in Mathematics & Statistics 536,
https://doi.org/10.1007/978-3-032-17666-0_9

Thus, it is essential to establish effective systems for monitoring and ensuring the quality of drinking water. In this context, the use of machine learning algorithms to assess and predict water quality is of paramount importance. These tools provide an innovative approach to analyzing and forecasting water quality, enabling policy-makers and water resource managers to make informed decisions to ensure a reliable supply of safe drinking water. This crisis is urgent, and several studies in the literature [4–7] have proposed solutions to address it.

In developing countries, rapid population growth is straining cities, leading to food insecurity and high unemployment. Urban agriculture, a fast-growing sector, presents a promising alternative for reducing youth unemployment and providing fresh produce to cities. In fact, 70% of the Dakar region's vegetable supply comes from urban agriculture, a trend also observed in other developing countries such as Mali and Ghana. However, this expansion is hindered by water shortages and limited land availability.

Approximately 70% of the world's freshwater, and 95% in developing countries, is used for agriculture. In response to water stress in arid and semi-arid regions, irrigation using wastewater has become increasingly common. Today, around 200 million farmers irrigate more than 20 million hectares with treated or untreated wastewater. However, this practice poses significant environmental risks, particularly the contamination of groundwater by microbiological and chemical agents.

Microbiological contamination of groundwater is a major public health issue, especially as it contributes to non-potable water contamination, which is responsible for 88% of diarrheal cases worldwide and 1.5 million deaths annually, mainly among children. In the Dakar region of Senegal, irrigation with polluted water in fragile ecosystems increases environmental and health risks, emphasizing the importance of real-time monitoring of the physicochemical parameters of both treated and untreated wastewater.

Currently, more than a billion people globally lack access to basic water supplies, and half of the population in developing countries suffers from diseases linked to water contamination. Water-related health risks are among the most severe and widespread global challenges. Each year, approximately 3.5 million people die due to issues related to water, sanitation, and hygiene, particularly in developing countries [8]. Diarrheal diseases, often caused by the consumption of contaminated water, claim the lives of over 1.5 million children under the age of five annually [9].

Nevertheless, providing adequate sanitation and access to safe drinking water could reduce diarrheal diseases by nearly 90% [10]. It is estimated that around 10% of global illnesses could be prevented through improved access to drinking water, sanitation, hygiene, and better environmental management.

Approximately 2.5 billion people live without access to improved sanitation, a major source of water pollution [11]. Unsafe or inadequate water consumption accounts for about 3.1% of global deaths. Infectious diseases, particularly those water-related, are the leading cause of death in children under the age of five [12]. Unsafe water contributes to 4 million cases of diarrheal disease and 2.2 million deaths annually [13]. Additionally, 1.3 million people die from malaria [14], and 160 million individuals are infected annually by schistosomiasis, a prevalent disease in Sub-Saharan Africa. Furthermore, 500 million people are at risk of contracting trachoma, while 1.5 million cases of hepatitis A occur each year [15]. In total, over 5 million people lose their lives annually due to water-related diseases [12].

Effective management and monitoring of wastewater quality are critical for protecting the environment and public health. In Senegal, existing infrastructures for treating and monitoring wastewater are often inadequate, leading to frequent resource contamination and public health issues.

Despite the importance of wastewater management, current monitoring methods in Senegal largely rely on manual inspections and laboratory analyses, which are often costly, slow, and limited in scope. These methods do not allow for real-time monitoring of the physicochemical and microbiological parameters of wastewater, nor do they support timely actions to prevent health and environmental risks. Therefore, there is an urgent need for a more efficient, automated monitoring system capable of providing real-time data on wastewater quality.

To address these challenges, we propose the development of an intelligent system based on the Internet of Things (IoT), Machine Learning (ML), and Deep Learning (DL) technologies. The objectives of this study are as follows:

- Develop an IoT sensor network capable of measuring the physicochemical parameters of wastewater in real time.
- Establish a communication structure to transmit collected data to a central analysis platform.
- Use ML and DL algorithms to analyze the collected data and quickly detect anomalies or contamination.

The integration of IoT into the fields of wastewater treatment and water quality prediction has the potential to revolutionize traditional approaches and tackle urgent challenges, especially given the global demand for clean water and sustainable systems [16].

This review provides an overview of the current state of wastewater management in Senegal, highlighting the challenges and impacts on public health. It also presents a survey of recent studies on online surface water monitoring systems, IoT, and machine learning technologies.

2 Case Study: Senegal

Wastewater management is a major issue in Senegal, a country facing environmental, health and socio-economic challenges linked to sanitation. As the urban population continues to grow, pressure on wastewater treatment infrastructures intensifies, highlighting the urgent need for sustainable, innovative solutions. Inadequate wastewater treatment has a direct impact on public health and the environment (Fig. 1).

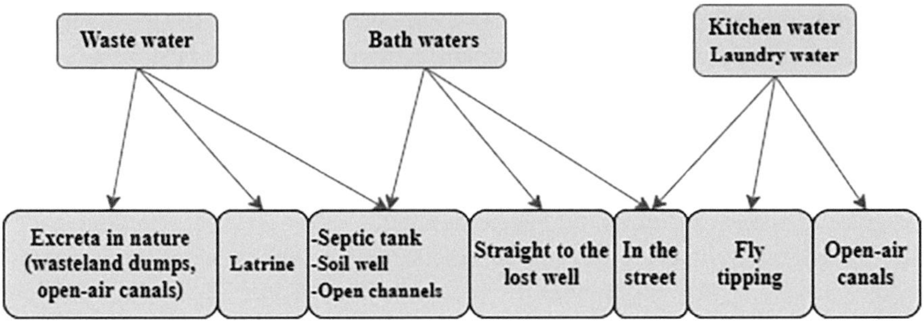

Fig. 1 Domestic wastewater management systems developed by populations in developing countries (example of Senegal) [17]

2.1 The Current State of Wastewater Management in Senegal

Urban and industrial growth in Senegal is putting pressure on water resources and contributing significantly to pollution problems. In developing countries such as Senegal, between 80 and 90% of wastewater, as well as around 70% of household and industrial waste, is discharged untreated into surface waters [10]. In Senegal, however, the main sources of wastewater are the domestic, industrial and agricultural sectors. In urban areas, sanitation is often rudimentary, with underdeveloped sewage systems, leading to frequent discharge of untreated wastewater into the environment. According to studies by Malick and Seydou [18], wastewater production in the Senegalese capital is around 67 million cubic meters (m^3) per year, and only a small quantity is treated in wastewater treatment plants. Furthermore, wastewater treatment infrastructures in Senegal are limited and concentrated mainly in large cities such as Dakar. However, these facilities are often undersized in relation to growing needs, and many rural communities have no access to treatment systems. Senegal currently has 13 wastewater treatment plants with a total treatment capacity of 36,066 m^3/day.

2.2 Wastewater Discharge Standards

Water is altered by human activity, whether domestic, industrial, artisanal, agricultural or other. Indeed, after use, water is said to be "polluted" and needs to be treated before it reaches the natural environment. Otherwise, it could cause serious damage [19]. Wastewater contains a variety of substances, including man-made contaminants such as chemicals, pathogens, nutrients, heavy metals and organic matter. However, wastewater must meet certain standards before it can be discharged. Table 1 shows Senegal's wastewater discharge standards.

Table 1 Wastewater discharge standards [20]

Parameters	Standards limits
Conductivity (μs/cm)	2700
pH	5.5–9.5
Dissolved oxygen (mg/l)	200
Temperature (T °C)	30
Turbidity (NTU)	5
DCO (mg/l)	200 if fd $\leq$ 100 kg/day 100
DBO (mg/l)	80 if fd $\leq$ 30 kg/day 40
Total nitrogen (mg/l)	30 if fd $\geq$ 100 kg/day
Phosphorus (P) (mg/l)	10 f fd $\geq$ 100 kg/day

2.3 Impacts of Wastewater in Senegal

In Senegal, the increasing discharge of inadequately treated wastewater, combined with agricultural runoff, has led to the degradation of groundwater and surface water quality in the Technopole area. The Senegalese capital discharged over 200,000 m^3 of wastewater per day in 2006 [21]. The consequences of these discharges can be categorized into three primary areas: (1) harmful effects on human health, (2) negative impacts on the environment, and (3) adverse repercussions on economic activities [22].

The uncontrolled discharge of wastewater has detrimental effects on aquatic ecosystems, including the pollution of rivers, lakes, and groundwater. This pollution contributes to the degradation of water quality, affecting biodiversity and drinking water resources.

Consumption of water contaminated by pathogens present in wastewater is a significant cause of waterborne diseases in Senegal. Several studies have highlighted the harmful consequences of using untreated or poorly treated water. Most pathogens (E. coli, Salmonella, Vibrio cholerae, Yersinia enterocolitica, Campylobacter jejuni), viruses (hepatitis A virus, enterovirus, rotavirus, Norwalk virus, etc.), and parasites (Giardia lamblia, Cryptosporidium parvum, Entamoeba histolytica, and Dracunculus medinensis) present in drinking water pose a serious risk of illness and must be prioritized for elimination [10].

The use of wastewater in agriculture (see Fig. 2) is a common practice due to water scarcity and the high concentration of fertilizers it contains. However, wastewater also contains large quantities and various forms of pathogenic microorganisms that can survive for varying periods in water, soil, or vegetables. These pathogens can contaminate and sicken farmers, traders, and consumers alike.

So, in Dakar, urban farmers water their crops with untreated wastewater, exposing themselves and consumers to significant health risks. Around 60% of these farmers suffer from intestinal parasitosis. Consumers of vegetables irrigated with this water

Fig. 2 Using wastewater in agriculture

are also at risk from the bacteria and viruses present. What's more, discharges pollute waters and coastlines, threatening public health and ecosystems, as well as food safety linked to seafood products [10].

3 Literary Review on Water Quality Monitoring

Aware of the high stakes involved in water quality, several organizations, research institutes, and companies have started to develop water quality monitoring systems to gain real-time insights into the parameters directly or indirectly influencing water. The Wapsmote Smart Water system from Libelium [23] is widely used for this purpose. The kit contains Wapsmote modules to which sensors are connected, and these modules transmit data to a Meshlium-type gateway [24] using ZigBee Pro. From the gateway, the data can be stored in a local network database or transmitted to a cloud database via the Internet or 3G/4G mobile networks. Researchers [25] have proposed a system that uses wireless sensor networks to remotely monitor water quality in real time. The system employs a reconfigurable sensor interface device with an FPGA board, sensors, a ZigBee-based wireless communication module, and a personal computer to monitor six water quality parameters. In 2015, [26] proposed an IoT-based system featuring an Internet of Things-based monitoring device.

In [27], the author proposed a model focused on creating an unmanned cruise ship for water quality monitoring. This model was effective for monitoring lake water quality, but it was also expensive and could not be used to monitor river water quality or for household and industrial applications. Additionally, the model failed under certain weather conditions and consumed a lot of energy. It was also unable to collect data from a specific control point in real time.

Some authors proposed a recurrent neural network model [28, 29] designed to predict lake water quality based on previously collected data. However, this model was not suitable for the task because water quality varies constantly, and it did not account for real-time data, rendering it ineffective for this purpose. Furthermore, it did not allow for real-time analysis of water quality. Similarly, another model was developed [30] for water quality prediction in lakes, using a gray-BP neural network model. However, this model was also inadequate for situations requiring real-time water quality monitoring.

In [31], the author proposed a BP network-based model for water quality assessment, which proved to be effective for assessing river water quality. However, the main limitation of this model was that it could not be used in different environments such as water treatment plants, lakes, and industrial sites. A similar model using the LSTM model [32] was proposed to analyze data for a specific river, such as the Fen or Yuncheng. In [33], Kageyama et al. proposed a water quality monitoring system tested on several sensors in a lake. Their system transmitted data every 10 min to a server via an Internet connection over the 3G mobile network, and the recorded data was displayed on a Google spreadsheet. However, they did not develop a platform for real-time visualization. Zulhani Rasin et al. used ZigBee-based technology with a compatible transceiver to implement a low-cost ad hoc system [34]. Sensors were deployed to establish a system designed to meet the monitoring requirements of the Water Framework Directive [35]. Research by N. Vijayakumar and R. Ramya focused on real-time water quality monitoring in an Internet of Things (IoT) environment. They monitored water quality using various sensors, such as pH, temperature, conductivity, dissolved oxygen, and turbidity sensors. These sensors were installed on a Raspberry Pi, with a keyboard and screen for reading data. To retrieve sensor data, the user had to issue an online command each time they wanted to see the readings. A Wi-Fi module connected to a mobile device was used to send the data to a server [36].

Another study by Mithila Barabde and Shruti Danve [37] on a real-time water quality monitoring system involved collecting sensor data and sending it to a server via SMS notifications. The sensors used in this research included pH, conductivity, and turbidity sensors. Each sensor was a node that transmitted data to a base station using radio frequency modules operating at 433 MHz. The sensor data was displayed on an LCD screen at the base station, and the data was sent back to the remote station using ZigBee modules. Using MAX232 serial data communication, the data was transferred to the server, and a notification was sent to users via SMS when the water content was in poor or dangerous condition using a GSM module [37].

4 Physicochemical Characteristics of Wastewater

Water quality refers to the assessment of various physical, chemical, biological, and microbiological parameters that determine the purity and safety of water for different uses, including human consumption, agriculture, industry, and the environment in general. Good water quality is crucial for protecting human health, supporting aquatic ecosystems, and preserving biodiversity.

4.1 Hydrogen Potential (pH)

It reflects the acidity or alkalinity of the environment. It is one of the key parameters in the process of eliminating pathogens and nutrients, as well as controlling odors. A pH that is too acidic inhibits the activity of methanogenic bacteria, resulting in unpleasant odors. However, a pH that is too basic reduces microbial activity [38]. pH can have a direct impact on human health and water quality in general. A pH that is too acidic or too basic can lead to unpleasant tastes and odors, making the water unpalatable. What's more, an inadequate pH can also influence the growth and reproduction of certain microbial organisms, such as bacteria, fungi and algae, which can cause disease [39, 40]. A high or low pH can also encourage the formation of limescale deposits or corrosion, which can lead to damage to water treatment and distribution equipment.

4.2 Turbidity (NTU)

Turbidity is an important parameter for assessing drinking water quality, as it measures the amount of suspended matter in the water. Suspended matter can be organic or inorganic particles, bacteria, viruses or algae [39, 40]. Drinking water with high turbidity can indicate the presence of contaminants potentially harmful to human health, such as pathogenic microorganisms, chemicals or heavy metals. In addition, turbid water can reduce the effectiveness of water disinfection treatments such as filtration and ultraviolet (UV) disinfection, as well as contaminants can protect pathogenic bacteria from treatment. Turbidity can be used to detect the presence of contaminant and assess the effectiveness of disinfection treatments.

4.3 Electrical Conductivity (EC)

Measures water's ability to conduct an electric current. This capacity is linked to the presence of salts dissolved in the water, such as sodium, calcium, magnesium, chloride ions, sulfate and lithium [39, 40]. The presence of dissolved salts in water can affect its taste, odor, color, and turbidity. The amount of salt dissolved in water is thus obtained by measuring conductivity. It is also a function of water temperature, and is more important as temperature rises [38].

4.4 Nitrate (NO^{3-})

The presence of nitrates in water is a major concern because of its implications for water quality and public health. Nitrates can originate mainly from the excessive use of nitrogen fertilizers in agriculture. At high levels in water, they can be a cause for concern, as they can be converted to nitrite in the body, which can interfere with the blood's ability to carry oxygen [39, 40]. This can be particularly problematic for infants, causing the condition known as methemoglobinemia or blue baby syndrome.

4.5 Dissolved Oxygen

Dissolved oxygen concentrations are important for characterizing water quality [38]. Oxygen is essential for all forms of aquatic life, including the organisms responsible for self-purification processes in natural waters. The oxygen content of natural waters varies with temperature, salinity, turbulence, the photosynthetic activity of algae and plants, and atmospheric pressure [41].

4.6 Suspended Solids (SS)

Suspended Solids (SS) are organic or mineral particles that are not dissolved in the effluent and remain in suspension. They represent a fraction of the pollution composed of these unsettled particles. According to aquaculture guidelines, the maximum recommended concentration of TSS is 50 mg/L, while the average observed in the lake is 85.9 mg/L. These high levels can reduce the penetration of sunlight, lower dissolved oxygen concentrations, and thus limit the development of aquatic life, going so far as to cause fish asphyxiation by clogging their gills [41].

4.7 Chemical Oxygen Demand (COD)

COD is a measure of the oxygen equivalent of organic matter in a sample of water sample liable to be oxidized by a powerful chemical oxidant, such as potassium dichromate [$K_2Cr_2O_7$] or potassium permanganate [$KMnO_4$]. Its limit value is 250 mg/L for water intended for aquaculture [41]. However, COD values that are too high could reduce the dissolved oxygen value through oxidation of organic matter.

4.8 *Biological Oxygen Demand (BOD)*

Biological oxygen demand BOD5 is the quantity of oxygen required to oxidize biodegradable organic matter (generally lower than the COD value). It is a measure of the organic pollution in water that can be biologically degraded. Biodegradable organic matter content is determined indirectly by measuring the amount of oxygen quantity of oxygen consumed, at 20 °C and in the dark, by aerobic bacteria in the sample. In a highly polluted environment, low BOD5 values may be linked to the presence of inhibiting toxic elements. Its limit value is 50 mg/L [41].

5 Internet of Things (IoT)

The term "Internet of Things" (IoT) refers to a network of physical objects capable of connecting to the Internet or other communications networks. These objects are usually equipped with software, hardware, or other technologies that enable them to facilitate analytical processes, such as environmental sensing. In the water sector, internet-connected systems incorporating pressure, flow, or water quality sensors are commonly used [16]. The IoT is ushering in a new technological era in which objects can communicate, calculate, and transform information at remarkable speed. This innovative technology has aroused the interest of many researchers, who have contributed their studies and research [42].

6 Machine Learning Algorithms

Computer systems that can learn from data without being explicitly programmed are called ML. This field, part of Artificial Intelligence (AI), focuses on the creation of algorithms and statistical models (see Fig. 3) [16].

The development of predictive models capable of making predictions or decisions based on data-driven learning, is the fundamental goal of ML. There are several machine learning algorithms in the literature, but some remain valid and stand out from the rest in terms of anomaly detection based on the frequency, rarity and proportion of anomalies in a data set. These algorithms are widely used for water quality monitoring and wastewater treatment.

6.1 *ANN*

Artificial Neural Networks (ANNs) are mathematical models designed to simulate the behavior of Biological Neural Networks (BNNs) in data processing. ANNs use node

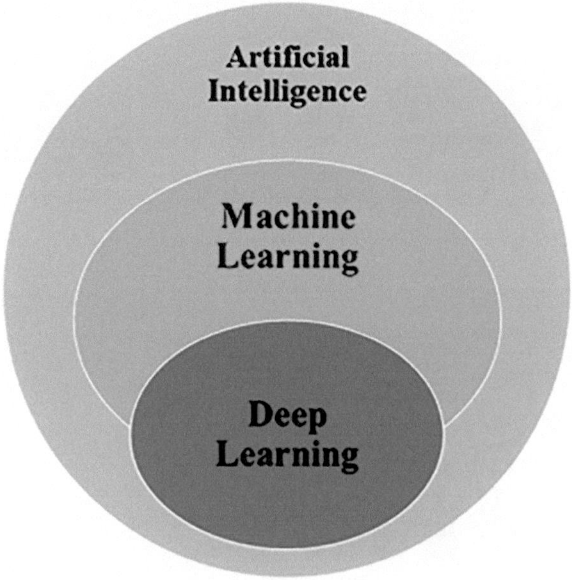

Fig. 3 Artificial intelligence methods

units to mimic neurons, performing information processing by adjusting the weights of interconnections between multiple nodes in the network. Typical components of an ANN include an input layer, an output layer, and several hidden layers located between them (see Fig. 4b). To perform complex non-linear calculations, ANNs use various activation functions, such as sigmoid, tanh, and ReLU functions, as well as several variable weights between neurons. Increasing the number of hidden layers in an ANN makes it possible to create more complex non-linear models and improve their expressive capabilities. ANNs can be trained to understand complex non-linear relationships between inputs and outputs by formulating and optimizing loss functions. The most commonly used ANN models include Recurrent Neural Networks (RNNs), CNNs, FNNs, and DNNs [16].

6.1.1 FNN

Fuzzy Neural Network (FNN) is a hybrid model combining ANNs and fuzzy logic to handle uncertain or ambiguous situations. It uses fuzzy logic to process input data, then an ANN to train and produce results (see Fig. 4d). The structure of the FNN is similar to that of a conventional neural network, but incorporates fuzzy logic to define the fuzzy relationship between inputs and outputs, as well as the weights of the connections between neurons (normalization, fuzzy inference and membership function). The FNN architecture comprises an input layer, a membership function layer, a fuzzy inference layer, a normalization layer and an output layer. FNN offers

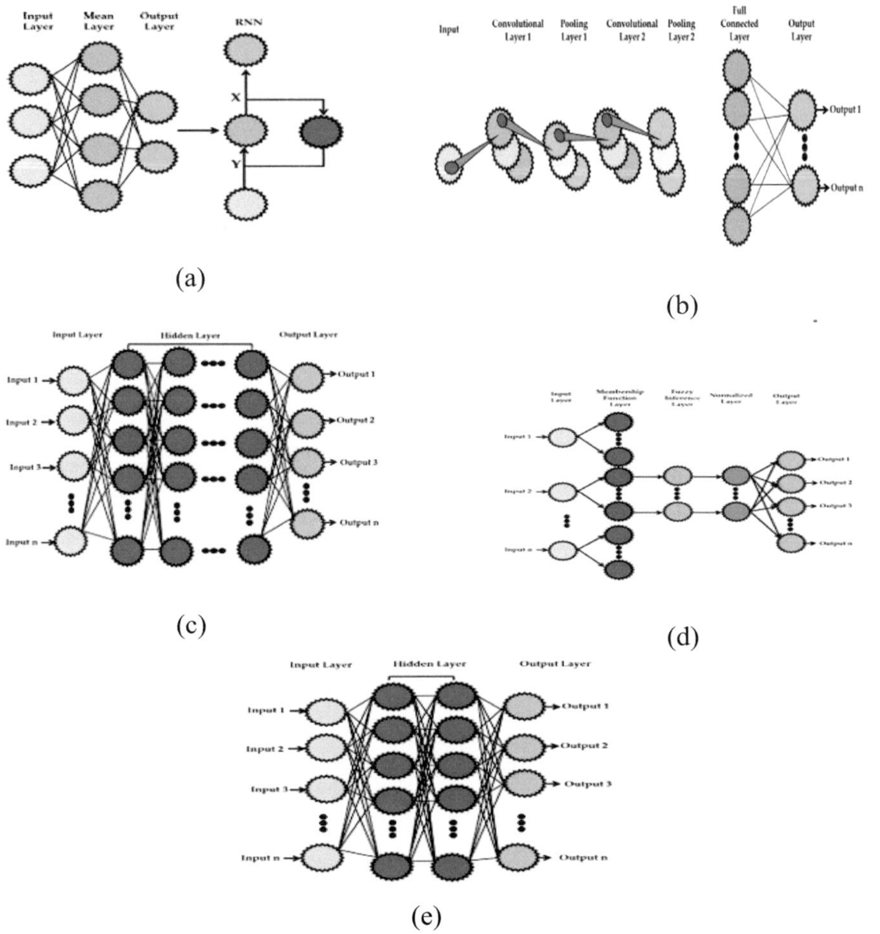

Fig. 4 Basic layout of ANN models. **b** ANN; **a** RNN; **c** FNN; **d** DNN; and **e** CNN

significant advantages over standard neural networks in the processing of complex problems, and is widely used in pattern recognition and control.

6.1.2 DNN

The Deep Neural Network (DNN) is a type of artificial neural network composed of several hidden layers located between the input and output layers. Its deep architecture enables it to learn hierarchical representations of data, combining low-level features in successive layers to learn higher-level features (see Fig. 4c). Like other neural networks, DNNs can contain a large number of hidden layers and neurons, making them suitable for capturing complex patterns in data or for learning highly non-linear correspondences between inputs and outputs. However, learning DNNs can be difficult and costly due to the complexity of the network architecture, requiring a large amount of data for training [16].

6.1.3 Recurrent Neural Network (RNN)

The RNN is a class of neural networks with recurrent connections, making it possible to process sequential data by integrating recursions as the sequence evolves. Thanks to its memory, parameter sharing and Turing completeness capabilities, the RNN excels at learning the non-linear characteristics of time series. The most commonly used type of RNN is Long Short-Term Memory (LSTM), which overcomes the gradient vanishing problem encountered in traditional RNNs by integrating additional gated units. RNN is proving particularly effective in applications involving water and wastewater treatment, water quality management and irrigated agriculture. Figure 3b illustrates a simple RNN architecture, where the output of the hidden layer is held in memory "W," which can be seen as an additional input for the RNN [43].

6.1.4 CNN

Convolutional Neural Networks (CNNs) are a deep learning technique that uses convolutional computation and a deep-structure feedforward neural network. CNNs are widely used in fields such as computer vision and natural language processing, among others. They employ convolutional layers to extract complex features from input images, pooling layers to reduce feature dimensionality, and fully connected layers for classification or regression tasks. The typical architecture of a CNN includes input, output, convolution, pooling, and fully connected layers (see Fig. 4b) [16].

7 Proposed Solution

As part of the implementation of the proposed smart system for wastewater quality monitoring, a pilot study is planned in the Technopole area of Dakar. This site was chosen due to its high environmental and health vulnerability linked to untreated domestic and industrial wastewater discharges, which are often used for irrigating vegetable crops. The main objective is to test the technical, economic, and functional feasibility of a real-time wastewater monitoring system based on a network of IoT sensors coupled with artificial intelligence algorithms for anomaly detection and risk prediction. The experimental setup will include the installation of 5–10 sensors at critical discharge points. These sensors will continuously measure essential physical and chemical parameters such as pH and COD. Each sensor will be connected to a microcontroller incorporating a LoRaWAN wireless communication module, enabling secure data transmission to a central gateway before being transferred to an analysis and visualization platform.

The collected data will be pre-processed and then analyzed using deep learning models to detect anomalies, predict drifts, and trigger automatic alerts if critical thresholds defined by Senegalese regulations are exceeded. An intuitive web interface will allow stakeholders to view the status of wastewater quality in real time. At the same time, manual samples will be taken regularly and analyzed in the laboratory to validate the accuracy of the measurements from smart sensors. This pilot phase will enable the relevance of the system to be assessed in a real-world context, identify the constraints associated with its large-scale deployment, and make the necessary adjustments before a possible transition to regional or national-scale smart, integrated, and sustainable water quality monitoring.

8 Conclusion

In conclusion, wastewater management in Senegal remains a major challenge, exacerbated by rapid urbanization, lack of adequate infrastructure, and often inadequate regulation. Although significant progress has been made, notably with the introduction of new technologies and participatory management models, many barriers remain. To move toward sustainable management, it is imperative to strengthen institutional capacities, improve public awareness, and promote investment in innovative solutions adapted to local realities. Collaboration between government players, non-governmental organizations, and local communities will be crucial to achieving these goals. Finally, ongoing research is needed to adapt and refine wastewater management strategies to the country's changing dynamics.

This study has provided an analysis and overview of various ML models, as well as advanced technologies and techniques applied in different water-related applications. AI and ML approaches have been used to improve, model, and automate processes in wastewater treatment, water-based agriculture, monitoring, and management of natural systems. The integration of AI/ML technologies is expected to reduce costs, improve water-related applications, and offer computer-aided solutions for complex water chemistry and physical/biological process challenges. ML and AI methods have been used to predict, model, automate and optimize important applications in water-related industries and operations, including water and wastewater treatment facilities, natural systems, and water-based agriculture.

Future work will focus on strengthening the resilience of water infrastructure by integrating AI/ML technologies that enable dynamic and predictive adaptation of monitoring systems. In addition, the design of innovative flexible sensors capable of being deployed in complex environments and detecting a wider range of contaminants is a priority. The methodologies envisaged include prototyping new materials for sensors, improving energy autonomy, and integrating continuous learning algorithms to refine the accuracy of predictions based on environmental changes, thereby paving the way for robust and sustainable intelligent systems.

References

1. United Nations Educational, Scientific and Cultural Organization (UNESCO).: UN-water report. UNESCO, Paris, France (2023)
2. United Nations Children's Fund (UNICEF).: For every child, reimagine. UNICEF Annual Report 2019. UNICEF Division of Communication, New York, NY, USA (2020)
3. United Nations Educational, Scientific and Cultural Organization (UNESCO).: UNESCO Report 2021. UNESCO, Paris, France (2023)
4. Ainapure, B., Baheti, N., Buch, J., Appasani, B., Vidyakant, J., Srinivasulu, Avireni.: Predicting drinking water potability using machine learning approaches: a case study of Indian rivers. Water Pract. Technol., (2023)
5. El Fadel, D.: Water resources evaluation and potability in North-East of Algeria, vol. 5. Nr. 4 2021, pp. 192–198, (2021)
6. Dorado-Guerra, D.Y., Corzo-P'erez, G., ParedesArquiola, J., P'erez-Mart'ın, M.A.: Machine learning models to predict nitrate concentration in a river basin. Environ. Res. Commun., **4**(12), 125012. IOP Publishing, (2023)
7. Shi, X., Cauquoin, A., Lohmann, G., Jonkers, L., Wang, Q., Yang, H., Sun, Y., Werner, M.: GMD Supplement.pdf, (2023)
8. OMS (ORGANISATION MONDIALE DE LA SANTE).: La charge mondiale de morbidite: Mise à jour 2004. Geneve, OMS (2008)
9. Black R.E., Cousens, S., Johnson, H.L., Lawn, J.E., Rudan, I., Bassani, D.G., Jha, P., Campbell, H., Walker, C.F., Cibulskis, R., Eisele, T., Liu, L., Mathers, C.: Presentation pour le groupe de reférence sur l' épidémiologie de la santé de l'enfant de l'OMS et de l'UNICEF. Les causes mondiales, regionales et nationales de la mortalité infantile en 2008: A systematic analysis. The Lancet 375(9730), 1969–1987 (2010)
10. Faye, C.: Les defis de la pollution de l'eau, une menace pour la sante publique: Atouts et defauts des lois et politiques de l'eau au Senegal. Larhyss J., **32**, 107–126 (2017)
11. UNICEF/WHO.: Programme commun de surveillance de l'approvisionnement en eau et de l'assainissement de l'UNICEF et de l'OMS. Progrès en matière d'eau potable et d'assainissement: Special Focus on Sanitation (Progres en matière d'eau potable et d'assainissement: accent particulier sur l'assainissement). UNICEF, New York et OMS, Geneve (2008)
12. Journee Mondiale De l'eau.: L'eau au service du développement. Le vendredi 22 mars 2002 est la Journee mondiale de l'eau, avec pour théme L'eau pour le developpement, (2002)
13. OMS, UNICEF.: Rapport sur l'evaluation mondiale de l'approvisionnement en eau et de l'assainissement, (2000)
14. OMS (ORGANISATION MONDIALE DE LA SANTE).: Eau, assainissement et hygiene, (2005)
15. OMS (Organisation Mondiale De La Sante).: Rapport sur la santé dans le monde: Changer l'histoire. Geneve (2004)
16. Alprol, A.E., Mansour, A.T., Ibrahim, M.E.E.-D., Ashour, M.: Artificial intelligence technologies revolutionizing wastewater treatment: current trends and future prospective. Water **16**, 314 (2024). https://doi.org/10.3390/w16020314
17. Bop, D., Gueye, M.T.: Gestion des eaux usees et des boues de vidange a dakar particulierement dans la zone humide du technopole, (2020).
18. Gaye, M., Niang, S.: Manual of good practices for the sound use of wastewater in urban agriculture. ENDA RUP (Relais pour le Developpement Urbain Populaire), Ouest Foire, BP 27 083, Malick Sy, Dakar, Senegal (2010)

19. Google. https://www.cieau.com/le-metier-de-leau/ressource-en-eau-eau-potable-eaux-usees/quest-ce-que-les-eaux-usees/. (2024)
20. EAUX USEES.: Normes de rejet, norme senegalaise NS 05–061, Juillet, (2001)
21. Akpo, Y.: Evaluation of the pollution of domestic wastewater collected and treated at the Camberéne wastewater treatment plant (Dakar), p. 2006. University of Dakar, Senegal, Memoire de diplôme d'études aprofondies de productions animales (2006)
22. Gueye, M.T., Bop, D., Ndoye, A., Sorlini, S.: Amelioration De La Qualité Des Eaux (Eaux Usées, Nappe Phréatique Et Lac) Par Phytoépuration Dans La Zone Du Technopole De Dakar (Sénégal) (2022)
23. DFROBOT.: Analog industrial ph sensor/meter pro kit v2. Datasheet, (2023)
24. Gaston, S., Fagnon, G.: Sysml: les diagrammes. Technologie, (2012)
25. Konde, S., Deosarkar, D.S.: Systeme de surveil-lance de la qualité de l'eau bas ´ e sur iot. (2020)
26. Lin, D.: Systeme de surveillance de la qualité de l'eau base sur IOT. (2015)
27. Wen, R., Shuo, J., Yonghui, Z., Kebin, T.: Le bateau de croisiere autonome sans pilote pour la surveillance et l'echantillonnage de la qualité de l'eau. In: Conference internationale sur les systèmes informatiques, l'électronique et le contrôle (ICCSEC), Dalian, pp. 700–703 (2017)
28. Wang, H., Gao, Y., Xu, Z., Xu, W.: An recurrent neural network application to forecasting the quality of water diversion in the water source of lake Taihu. In: International conference on remote sensing, environment and transportation engineering, Nanjing, pp. 984–988 (2011)
29. Zhao, Y., Zou, Z., Wang, S.: A back propagation neural network model based on kalman filter for water quality prediction (Un modèle de reseau neuronal à propagation arrière basé sur le filtre de Kalman pour la prediction de la qualité de l'eau). In: 1th international conference on natural computation (ICNC), pp. 149–153. IEEE, (2015)
30. Wang, X. et al.: The lake water bloom intelligent prediction method and water quality remote monitoring system. In: Sixth international conference on natural computation, Yantai, pp. 3443–3446 (2010)
31. Zhang, Y.Y., Feng, M.Q., Hao, Z.L.: Water quality assessment based on bp network and its application. In: 2011 International symposium on water resource and environmental protection, vol. 2, pp. 872–876. IEEE, (2011)
32. Wang, Y., Zhou, J., Chen, K., Wang, Y., Liu, L.: Water quality prediction method based on LSTM neural network. In: 2017 12th International Conference on Intelligent Systems and Knowledge Engineering (ISKE), pp. 1–5. IEEE, (2017)
33. Miura, M., Maeda, A., Kageyama, T.: A wireless sensor network platform for water quality monitoring (Une plateforme de reseau de capteurs sans fil pour la surveillance de la qualité de l'eau), pp. 1–3. IEEE, (2016)
34. Rasin, Z., Abdullah M.R.: Water quality monitoring system using zigbee based wireless sensor network journal. Int. J. Eng. Technol., **9**(10), 24–28 (2009)
35. Martinez, R., Cleary, J., Slater, C., Regan, F., Diamond, D., Murphy, H., O'Flyrm, B.: Smart-coast: a wireless sensor network for water quality monitoring. In: 32nd IEEE conference on, pp 815–816. IEEE, (2007)
36. Vijayakumar, N., Ramya, R.: La surveillance en temps réel de la qualité de l'eau dans l'environnement Iot. Int. J. Sci. Res. (IJSR), **4**(3), (2015)
37. Barabde, M.N., Danve, S.R.: Systeme de surveillance continue de la qualité de l'eau pour les ressources en eau dans les endroits éloignés. Int. J. Eng. Res. Gen. Sci. **3**, Part 2, (2015)
38. Rim Yêgb Nonma Valérie.: Zongo, Evaluation des performances epuratoires de la station d'epuration des eaux usees de kossodo apres 9 ans de fonctionnement, (2013)
39. World Health Organization (WHO).: Guidelines for drinking-water quality: 4th ed. Incorporating First Addendum. (2017)
40. World Health Organization (WHO).: Guidelines for drinking-water quality: 4th ed. Incorporating First and Second Addenda. (2017)
41. Gueye, M.T., Bop, D., Ndoye, A., Sorlini, S.: Impacts de la pollution du lac dans la zone du technopole de Pikine sur le risque de disparition des Especes aquatiques

42. Mouha, R.A.: Internet des objets (IoT). J. Data Anal. Inf. Process. **9**, 77–101 (2021). https://doi.org/10.4236/jdaip.2021.92006
43. Wang, Y., Cheng, Y., Liu, H., Guo, Q., Dai, C., Zhao, M., Liu, D.: Examen des applications de l'intelligence artificielle dans le traitement des eaux usées. Sustainability. **15**, 13557 (2023). https://doi.org/10.3390/su151813557

On the Averaging Principle for Multidimensional BSDEs Driven by Fractional Brownian Motion

Djibril Ndiaye and Yaya Sagna

Abstract This paper is devoted to presenting an averaging principle for multidimensional backward stochastic differential equations driven by fractional Brownian motion (MFrBSDEs for short). Under a Lipschitz condition, the solutions to MFrBSDEs can be approximated by solutions to averaged stochastic systems in the mean-square sense and probability. Moreover, our results have significantly generalized some previous work.

Keywords Averaging principle · Backward stochastic differential equation · Stochastic calculus · Fractional brownian motion · Chebyshev's inequality and itô's representation formula

AMS Subject Classification 60H05 · 60G44

1 Introduction

An existence and uniqueness result for the solution of multidimensional backward stochastic differential equations (MBSDEs in short) on finite time interval was first established by Pardoux and Peng [12] under an uniform Lipschitz condition on the generator. Since then, BSDEs have been successfully applied in various areas such as mathematical finance, stochastic control, stochastic games, partial differential equations and so on.

Many efforts have been made to weaken the Lipschitz condition on the coefficient. To the best of our knowledge, Mao [8] successfully investigated the one dimensional case of these equations with the so-called Osgood condition on the generator. Latter,

D. Ndiaye (✉) · Y. Sagna
Laboratoire de Mathématiques Appliquées, Université Cheikh Anta Diop De Dakar, Dakar-Fann Sénégal, Dakar, Senegal
e-mail: djibril1.ndiaye@ucad.edu.sn

Y. Sagna
e-mail: yaya7.sagna@ucad.edu.sn

© The Author(s), under exclusive license to Springer Nature Switzerland AG 2026
C. T. Gueye et al. (eds.), *Mathematics, Computer Science, Cybersecurity and Artificial Intelligence*, Springer Proceedings in Mathematics & Statistics 536,
https://doi.org/10.1007/978-3-032-17666-0_10

Fan et al. [5] generalize the result of Mao [8] to MBSDEs under the Osgood condition, and the author extended in [4] the existence, uniqueness, and stability of L^1 solutions for MBSDEs with generators of one-sided osgood type.

As a generalization of BSDEs, several authors have investigated BSDEs with respect to fractional Brownian motion $(B^H(t))_{t\geq 0}$ with Hurst parameter H. Thanks to its self-similarity property (i.e., $B^H(at)$ has the same law as $a^H B^H(t)$ for any $a > 0$), this process is a powerful tool in studying models of physical phenomena and finance that exploit the self-similarity and the long-range dependence.

Bender [2] gave one of the earliest results on fractional BSDEs. He established an explicit solution of a class of linear fractional BSDEs with arbitrary Hurst parameter H. This was achieved essentially by means of solution of a specific linear parabolic PDE. There are two major obstacles due to the properties of fractional Brownian motion.

Firstly, the fractional Brownian motion is not a semimartingale when $H \neq 1/2$, so the classical Itô calculus based on semimartingales cannot be directly transposed to the fractional case. Second, there is no martingale representation theorem with respect to the fractional Brownian motion. Hu and Peng [6] successfully overcame this problem in the case $H > 1/2$ by means of the quasi-conditional expectation. The authors prove existence and uniqueness of the solution but with under restrictive assumptions on the generator.

Recently, Aidara et al. [1] established an averaging principle for BSDEs driven by two mutually independent fractional Brownian motions.

Our goal in this work is to extend the result obtained in [1] to MFrBSDEs whose generator satisfies the Lipschitz condition. The paper is organized as follows. In Sect. 2, we recall some basic notions on fractional calculus for multidimensional fractional Brownian motion. In Sect. 3, we establish our main result.

2 Basic Notions on Fractional Calculus for Multidimensional Fractional Brownian Motion

In this section, we briefly recall some basic facts concerning stochastic calculus with respect to a multidimensional fractional Brownian motion; we refer to [11] for further details. Let us consider m independent fractional Brownian motions $B^{H_1}, B^{H_2}, \ldots, B^{H_m}$ with respectively Hurst parameters $H_1, H_2, \ldots, H_m$, where $H_k \in (\frac{1}{2}, 1)$, $k = 1, 2, \ldots, m$.

Let Ω be a non-empty set, $\mathcal{F}$ a σ-algebra of sets on Ω, $\mathbb{P}$ a probability measure defined on $\mathcal{F}$.

Let $\{\mathcal{F}_t, t \in [0, T]\}$ be the filtration generated by the m fractional Brownian motions.

The triplet $(\Omega, \mathcal{F}, \mathbb{P})$ defines a probability space and $\mathbb{E}$ denotes the mathematical expectation with respect to the probability measure $\mathbb{P}$.

It is known that $\mathbb{E}(B_t^{H_k}) = 0$ and $\mathbb{E}\left(B_t^{H_k}.B_s^{H_j}\right) = \frac{1}{2}\left(|t|^{2H_k} + |s|^{2H_k} - |t-s|^{2H_k}\right)$ δ_{kj} where δ_{kj} denotes the Kronecker's delta : for $1 \leq k, j \leq m$, $\delta_{kj} = 1$ if $k = j$ and $\delta_{kj} = 0$ if otherwise.

Now, fix a Hurst index $H_k \in (\frac{1}{2}, 1)$ and define

$$\phi_k(x) = H_k(2H_k - 1)|x|^{2H_k - 2}, \quad x \in \mathbb{R}.$$

Let us denote by $|\mathcal{H}_k|$ the Banach space of measurable functions $f : [0, T] \to \mathbb{R}$ such that

$$\|f\|^2_{\mathcal{H}_k} := \int_0^T \int_0^T \phi_k(u - v)|f(u)||f(v)|dudv < \infty.$$

Given $\xi, \eta \in |\mathcal{H}_k|$, we define the quantity

$$\langle \xi, \eta \rangle_{k,T} = \int_0^T \int_0^T \phi_k(u - v)\xi(u)\eta(v)dudv.$$

For $\xi = \eta$,

$$\|\xi\|^2_{k,T} = \langle \xi, \xi \rangle_{k,T} = \int_0^T \int_0^T \phi_k(u - v)\xi(u)\xi(v)dudv < \infty.$$

Then $\langle \xi, \eta \rangle_{k,T}$ defines a Hilbert scalar product. Let $\mathcal{H}_k$ be the completion of the space of step functions in $|\mathcal{H}_k|$ under this scalar product. We emphasize that the elements of $\mathcal{H}_k$ can be distributions. Moreover, from [9] there is a the continuous embedding

$$L^2([0, T]) \subset L^{\frac{1}{H_k}} \subset |\mathcal{H}_k| \subset \mathcal{H}_k.$$

Let $\mathcal{P}_T^{H_k}$ denote the set of all polynomials of the fractional Brownian motion $(B_t^{H_k})_{t\geq 0}$. Namely, $\mathcal{P}_T^{H_k}$ contains all elements of the form:

$$F(\omega) = f\left(\int_0^T \xi_1(t)dB_t^{H_k}, \ldots, \int_0^T \xi_l(t)dB_t^{H_k}\right),$$

where f is a polynomial function of l variables and $\xi_1, \ldots, \xi_l \in \mathcal{H}_k$.

The Malliavin derivative operator D^{H_k} of F is defined as follows:

$$D_s^{H_k} F = \sum_{i=1}^{l} \frac{\partial f}{\partial x_i}\left(\int_0^T \xi_1(t)dB_t^{H_k}, \ldots, \int_0^T \xi_l(t)dB_t^{H_k}\right)\xi_i(s), \quad s \in [0, T].$$

In the sequel, let us introduce another derivative, called the weighted Malliavin ϕ_k-derivative of F denoted by $\mathbb{D}_t^{H_k} F$ defined as

$$\mathbb{D}_t^{H_k} F = \int_0^T \phi_k(t-s) D_s^{H_k} F ds,$$

where ϕ_k is suitable kernel function. We denote by $\mathbb{L}_{H_k}^{1,2}$ the space of all stochastic processes $F : (\Omega, \mathcal{F}, \mathbb{P}) \to \mathcal{H}$ such that

$$\mathbb{E}\left(\|F\|_{k,T}^2 + \int_0^T \int_0^T |D_s^{H_k} F_t|^2 ds dt \right) < \infty.$$

Theorem 2.1 *Let $F : (\Omega, \mathcal{F}, \mathcal{P}) \to \mathcal{H}$ be a stochastic process such that*

$$\mathbb{E}\left(\|F\|_{k,T}^2 + \int_0^T \int_0^T |D_s^{H_k} F_t|^2 ds dt \right) < \infty.$$

Then, the Itô-type stochastic integral $\int_0^T F_s dB_s^{H_k}$ exists in $L^2(\Omega, \mathcal{F})$.

Moreover, we have

$$\mathbb{E}\left(\int_0^T F_s dB_s^{H_k} \right) = 0$$

and

$$\mathbb{E}\left[\left(\int_0^T F_s dB_s^{H_k} \right)^2 \right] = \mathbb{E}\left(\|F\|_{k,T}^2 + \int_0^T \int_0^T D_s^{H_k} F_t D_t^{H_k} F_s ds dt \right).$$

The following lemma will be very useful in the implementation of our stochastic averaging principle. For the proof, we refer the reader to [13].

Lemma 2.2 *Let $k \in \{1, \ldots, m\}$ and $B_t^{H_k}$ be a fractional Brownian motion with Hurst parameter $H_k \geq \frac{1}{2}$. Let $u(s)$ be a stochastic process in the space $\mathbb{L}_{H_k}^{1,2}$. Then, for every $T < \infty$, there exists a constant C_0 such that*

$$\mathbb{E}\left[\left(\int_0^T |u(s)| dB_s^{H_k} \right)^2 \right] \leq 2H_k T^{2H_k - 1} \mathbb{E}\left(\int_0^T |u(s)|^2 ds \right) + C_0 T^2.$$

Theorem 2.3 (Multidimensional Fractional Itô's Formula) *Let $\sigma_{ik} \in L^2([0, T])$, $i = 1, \ldots, d$, $k = 1, \ldots, m$ be some deterministic functions. Assume that the scalar products $\langle \sigma_{ik}, \sigma_{jk} \rangle_{k,t}$ are continuously differentiable functions in $t \in [0, T]$. Let $X_t = (X_t^1, \ldots, X_t^d)$ be defined by*

$$X_t^i = X_0^i + \int_0^t b_i(s) ds + \sum_{k=1}^m \int_0^t \sigma_{ik}(s) dB_s^{H_k}, \quad t \in [0, T], \quad i = 1, \ldots, d,$$

where, $X_0 = (X_0^1, \dots, X_0^d)$ is a constant vector and $b_i : [0, T] \to \mathbb{R}$ are deterministic continuous functions satisfying $\int_0^T |b_i(s)|ds < \infty$, for all i. Let F be continuously differentiable with respect to t and twice continuously differentiable with respect to x. Then:

$$\begin{aligned} F(t, X_t) &= F(0, X_0) + \int_0^t \frac{\partial F}{\partial s}(s, X_s)ds + \sum_{i=1}^d \int_0^t \frac{\partial F}{\partial x_i}(s, X_s)dX_s^i \\ &+ \sum_{i,j=1}^d \int_0^t \frac{\partial^2 F}{\partial x_i \partial x_j}(s, X_s) \sum_{k=1}^m \sigma_{ik}(s) D_s^{H_k}(X_s^j)ds \\ &= F(0, X_0) + \int_0^t \frac{\partial F}{\partial s}(s, X_s)ds + \sum_{i=1}^d \int_0^t \frac{\partial F}{\partial x_i}(s, X_s)b_i(s)ds \\ &+ \sum_{i=1}^d \sum_{k=1}^m \int_0^t \frac{\partial F}{\partial x_i}(s, X_s)\sigma_{ik}(s)dB_s^{H_k} \\ &+ \frac{1}{2} \sum_{i,j=1}^d \int_0^t \frac{\partial^2 F}{\partial x_i \partial x_j}(s, X_s) \sum_{k=1}^m \frac{d}{ds}(\langle \sigma_{ik}, \sigma_{jk} \rangle_{k,s})ds, \quad t \in [0, T]. \end{aligned}$$

Theorem 2.4 (Fractional Itô's Chain Rule) *Let $T \in (0, \infty)$ and let $b_1(s), b_2(s)$, $\sigma_{1k}(s), \sigma_{2k}(s)$ be in $\mathbb{D}_{1,2}$, the Malliavin space. Suppose*

$$\mathbb{E}\left(\int_0^T (|b_i(s)| + |\sigma_{ij}(s)|)ds \right) < \infty, \quad i = 1, 2, \ j = 1, \dots, m.$$

Assume that $\mathbb{D}_t^{H_j} \sigma_{ij}(s)$ are continuously differentiable with respect to $(s, t) \in [0, T]^2$ for almost all $\omega \in \Omega$ and satisfy

$$\mathbb{E}\left(\int_0^T \int_0^T |\mathbb{D}_t^{H_j} \sigma_{ij}(s)|^2 dsdt \right) < \infty, \quad i = 1, 2, \ j = 1, \dots, m.$$

Denote the processes

$$F(t) = F(0) + \int_0^t b_1(s)ds + \sum_{k=1}^m \int_0^t \sigma_{1k}(s)dB_s^{H_k}, \quad t \in [0, T]$$

and

$$G(t) = G(0) + \int_0^t b_2(s)ds + \sum_{k=1}^m \int_0^t \sigma_{2k}(s)dB_s^{H_k}, \quad t \in [0, T].$$

Then, the product $F(t).G(t)$ satisfies

$$
\begin{aligned}
F(t)G(t) &= F(0)G(0) + \int_0^t F(s)dG(s) + \int_0^t G(s)dF(s) \\
&\quad + \sum_{k=1}^m \left(\int_0^t \mathbb{D}_s^{H_k} F(s)\sigma_{2k}(s)ds + \int_0^t \mathbb{D}_s^{H_k} G(s)\sigma_{1k}(s)ds \right) \\
&= F(0)G(0) + \int_0^t F(s)b_2(s)ds + \sum_{k=1}^m \int_0^t F(s)\sigma_{2k}(s)dB_s^{H_k} \\
&\quad + \int_0^t G(s)b_1(s)ds + \sum_{k=1}^m \int_0^t G(s)\sigma_{1k}(s)dB_s^{H_k} \\
&\quad + \sum_{k=1}^m \left(\int_0^t \mathbb{D}_s^{H_k} F(s)\sigma_{2k}(s)ds + \int_0^t \mathbb{D}_s^{H_k} G(s)\sigma_{1k}(s)ds \right).
\end{aligned}
$$

3 Averaging Principle for SMFrBSDEs

3.1 BSDEs Driven by Multidimensional Fractional Brownian Motions

Assume that (H_1) $\sigma_{ik} : [0, T] \to \mathbb{R}$, $i = 1, \dots, d$, $k = 1, \dots, m$ are deterministic continuous, differentiable functions such that $\sigma_{ik}(t) \neq 0$, for all $t \in [0, T]$. Given a constant vector $\eta_0 = (\eta_0^1, \dots, \eta_0^d)$, define the process:

$$
\eta_t^i = \eta_0^i + \sum_{k=1}^m \int_0^t \sigma_{ik}(s)dB_s^{H_k}, \quad t \in [0, T],\ i = 1, \dots, m.
$$

Since

$$
\langle \sigma_{ik}, \sigma_{jk} \rangle_{k,t} = H_k(2H_k - 1) \int_0^t \int_0^t |v - u|^{2H_k - 2} \sigma_{ik}(v)\sigma_{jk}(u)(u)dvdu,
$$

we obtain

$$
\frac{d}{dt}(\langle \sigma_{ik}, \sigma_{jk} \rangle_{k,t}) = 2H_k(2H_k - 1)\sigma_{ik}(t) \int_0^t |t - u|^{2H_k - 2} \sigma_{jk}(u)du = 2\sigma_{ik}(t)\widehat{\sigma}_{jk}(t) > 0,
$$

where

$$
\widehat{\sigma}_{jk}(t) = \int_0^t \phi_k(t - u)\sigma_{jk}(u)du.
$$

We suppose: (H_2) $\xi = h(\eta_T)$ for some function h with bounded derivative and such that $\mathbb{E}(|\xi|^2) < \infty$,
(H_3) $f : \Omega \times [0, T] \times \mathbb{R}^d \times \mathbb{R} \times \mathbb{R}^m \to \mathbb{R}$ is a continuous function such that there exists a positive constant L satisfying for all $t \in [0, T]$, $y, y' \in \mathbb{R}$, $z, z' \in \mathbb{R}^m$,

$$|f(t, x, y, z) - f(t, x, y', z')| \leq L(|y - y'| + \|z - z'\|)$$

and

$$\mathbb{E}\left(\int_0^T |f(t, \eta_t, 0, 0)|^2 dt\right) < \infty.$$

Now, we consider the MFrBSDEs:

$$Y_t = \xi + \int_t^T f(s, \eta_s, Y_s, Z_s)ds - \sum_{k=1}^m \int_t^T Z_s^k dB_s^{H_k}, \quad 0 \leq t \leq T. \tag{1}$$

We introduce the following functional spaces:

- $\mathcal{C}^{1,3}_{\text{pol}}([0, T] \times \mathbb{R})$ is the space of all $\mathcal{C}^{1,3}$-functions over $[0, T] \times \mathbb{R}$, which together with their derivatives are of polynomial growth.
- $\mathcal{V}_T = \{Y = \Psi(., \eta) : \Psi \in \mathcal{C}^{1,3}_{\text{pol}}([0, T] \times \mathbb{R})$ and $\dfrac{\partial \Psi}{\partial t}$ is bounded, $t \in [0, T]\}$.
- $\widetilde{\mathcal{V}}_T$ the completion of $\succeq_T$ under the following norm:

$$\|Y\|_H^2 = \mathbb{E}\left(\int_0^T t^{2H-1}|Y|_t^2 dt\right) = \mathbb{E}\left(\int_0^T t^{2H-1}|\Psi(t, \eta_t)|^2 dt\right).$$

Definition 3.1 A pair of processes $(Y_t, Z_t)_{0 \leq t \leq T}$ is called a solution to MFrBSDE (1) if it satisfies Eq. (1) and such that $Y \in \widetilde{\mathcal{V}}_T^{H_1} \cap \widetilde{\mathcal{V}}_T^{H_2} \cap \ldots \cap \widetilde{\mathcal{V}}_T^{H_m}$ and $Z = (Z^1, \ldots, Z^k)$, where $Z^k \in \widetilde{\mathcal{V}}_T^{H_1 + H_k - 1/2} \cap \widetilde{\mathcal{V}}_T^{H_2 + H_k - 1/2} \cap \ldots \cap \widetilde{\mathcal{V}}_T^{H_m + H_k - 1/2}$.

Proposition 3.2 *Assume* (H_1)*-*(H_4)*. Then, there exists a unique solution of (1).*

The following proposition will play a crucial role in the sequel. For the proof we refer to [6, 9].

Proposition 3.3 *Let* (Y, Z) *be a solution of the MFrBSDE (1). Then for almost* $t \in [0, T]$*, we have*

$$\mathbb{D}_s^{H_k} Y_s = \int_0^T \phi(s - r) D_r^{H_r} Y_r dr = \frac{\widehat{\sigma}_k(s)}{\sigma_k(s)} Z_s^k. \tag{2}$$

Moreover, there exists $M_k > 0$ *such that for all* $s \in [0, T]$*,*

$$\frac{s^{2H_k-1}}{M_k} \leq \frac{\widehat{\sigma}_k(s)}{\sigma_k(s)} \leq M_k s^{2H_k-1}. \tag{3}$$

3.2 Averaging Principle

In this section, we focus on the averaging principle for the MFrBSDEs under Lipschitz coefficients. Let us consider the standard form of Eq. (1): for all $t \in [0, T]$

$$Y_t^\varepsilon = \xi + \varepsilon^{2H} \int_t^T f(r, \eta_r^\varepsilon, Y_r^\varepsilon, Z_r^\varepsilon) dr - \varepsilon^H \sum_{k=1}^m \int_t^T Z_r^{k,\varepsilon} dB_r^{H_k} \tag{4}$$

where $\eta_t^{i,\varepsilon} = \eta_0^i + \varepsilon^H \sum_{k=1}^m \int_0^t \sigma_{ik}(s) dB_s^{H_k}, \quad t \in [0, T], \, i = 1, \ldots, m,$ and $H = \min\{H_1, H_2, \ldots, H_m\}$.

By the previous results, Eq. (4) has an unique adapted and square integrable solution. We will see how the solution Y_t^ε can be approximated to the solution process $\overline{Y}_t$ of the simplified equation: for all $t \in [0, T]$

$$\overline{Y}_t = \xi + \varepsilon^{2H} \int_t^T \overline{f}(\eta_r^\varepsilon, \overline{Y}_r^\varepsilon, \overline{Z}_r^\varepsilon) dr - \varepsilon^H \sum_{k=1}^m \int_t^T \overline{Z}_r^{k,\varepsilon} dB_r^{H_k}, \tag{5}$$

where $(\overline{Y}_t, \overline{Z}_t)$ has the same properties as $(Y_t^\varepsilon, Z_t^\varepsilon)$.

We assume that the coefficients f and $\overline{f}$ are continuous functions and satisfy the following assumptions:

- (A_1) There exists $L > 0$ such that for all $(t, x, y, z, \tilde{y}, \tilde{z}) \in [0, T] \times \mathbb{R}^d \times \mathbb{R} \times \mathbb{R}^m \times \mathbb{R} \times \mathbb{R}^m$, we have

$$|f(t, x, y, z) - f(t, x, \tilde{y}, \tilde{z})|^2 \leq L \left(|y - \tilde{y}|^2 + \|z - \tilde{z}\|^2\right).$$

- (A_2) For any $t \in [0, T_1] \subset [0, T]$ and for all $(x, y, z) \in \mathbb{R}^d \times \mathbb{R} \times \mathbb{R}^m$, we have

$$\frac{1}{T_1 - t} \int_t^{T_1} |f(s, x, y, z) - \overline{f}(x, y, z)|^2 ds \leq \varphi(T_1 - t) \left(1 + |y|^2 + \|z\|^2\right)$$

where φ is a bounded function.

The lemma that we are going to state will also play an essential role in the rest of the work.

Lemma 3.4 *Suppose that the original SMFrBSDEs (4) and the averaged one (5) satisfy assumptions* (A_1) *and* (A_2). *For a given arbitrarily small number* $u \in [0, t] \subset [0, T]$, *there exist two constants* $L_1 > 0$ *and* $C_1 > 0$ *such that*

$$\mathbb{E}\left(\sum_{k=1}^{m}\int_u^T s^{2H_k-1}|Z_s^{k,\varepsilon}-\overline{Z}_s^k|^2ds\right)\leq L_1\mathbb{E}\left(\int_u^T |Y_s^\varepsilon-\overline{Y}_s|^2ds\right)+C_1(T-u). \tag{6}$$

Proof To simplify the notations, let us introduce $\overline{\Delta Y}^\varepsilon = Y^\varepsilon - \overline{Y}$ and $\overline{\Delta Z}^\varepsilon = Z^\varepsilon - \overline{Z}$.

It is obvious that the pair of processes $(\overline{\Delta Y}_t^\varepsilon, \overline{\Delta Z}_t^\varepsilon)_{0\leq t\leq T}$ solves the SMFrBSDE

$$\overline{\Delta Y}_t^\varepsilon = \varepsilon^{2H}\int_t^T [f(s,\eta_s^\varepsilon,Y_s^\varepsilon,Z_s^\varepsilon)-\overline{f}(\eta_s^\varepsilon,\overline{Y}_s,\overline{Z}_s)]ds-\varepsilon^H\sum_{k=1}^{m}\int_t^T\overline{\Delta Z}_s^{k,\varepsilon}dB_s^{H_k}.$$

By applying Itô's formula to $|\overline{\Delta Y}_t^\varepsilon|^2$, we get

$$\begin{aligned}|\overline{\Delta Y}_t^\varepsilon|^2 &= 2\varepsilon^{2H}\int_u^T \overline{\Delta Y}_s^\varepsilon[f(s,\eta_s^\varepsilon,Y_s^\varepsilon,Z_s^\varepsilon)-\overline{f}(\eta_s^\varepsilon,\overline{Y}_s,\overline{Z}_s)]ds\\ &\quad -2\varepsilon^H\sum_{k=1}^{m}\int_u^T\overline{\Delta Y}_s^\varepsilon\overline{\Delta Z}_s^{k,\varepsilon}dB_s^{H_k}-2\varepsilon^H\sum_{k=1}^{m}\int_u^T\mathbb{D}_s^{H_k}\overline{\Delta Y}_s^\varepsilon\overline{\Delta Z}_s^{k,\varepsilon}ds. \end{aligned}\tag{7}$$

Since the pair of processes $(\overline{\Delta Y}_s^\varepsilon, \overline{\Delta Z}_s^\varepsilon)_{t\leq s\leq T}$ is such that $\overline{\Delta Y}^\varepsilon \in \widetilde{\mathcal{V}}_T^{H_1}\cap\widetilde{\mathcal{V}}_T^{H_2}\cap\ldots\cap\widetilde{\mathcal{V}}_T^{H_m}$ and $\overline{\Delta Z}^\varepsilon = (Z^{1,\varepsilon}-\overline{Z}^1,\ldots,Z^{k,\varepsilon}-\overline{Z}^k)$, where

$$(Z^{k,\varepsilon}-\overline{Z}^k)\in\widetilde{\mathcal{V}}_T^{H_1+H_k-1/2}\cap\widetilde{\mathcal{V}}_T^{H_2+H_k-1/2}\cap\ldots\cap\widetilde{\mathcal{V}}_T^{H_m+H_k-1/2},$$

we have $\overline{\Delta Y}_s^\varepsilon\overline{\Delta Z}_s^{k,\varepsilon}\in\mathbb{L}_{H_k}^{1,2}$, $k=1,\ldots,m$. Then, by Theorem 2.1

$$\mathbb{E}\left(\int_0^T\overline{\Delta Y}_s^\varepsilon\overline{\Delta Z}_s^{k,\varepsilon}dB_s^{H_k}\right)=0.$$

We arrive at

$$\begin{aligned}\mathbb{E}\left(|\overline{\Delta Y}_t^\varepsilon|^2\right)&+2\varepsilon^H\mathbb{E}\left(\sum_{k=1}^{m}\int_u^T\mathbb{D}_s^{H_k}\overline{\Delta Y}_s^\varepsilon\overline{\Delta Z}_s^{k,\varepsilon}ds\right)\\ &=2\varepsilon^{2H}\mathbb{E}\left(\int_u^T\overline{\Delta Y}_s^\varepsilon[f(s,\eta_s^\varepsilon,Y_s^\varepsilon,Z_s^\varepsilon)-\overline{f}(\eta_s^\varepsilon,\overline{Y}_s,\overline{Z}_s)]ds\right)\\ &\leq 2\varepsilon^{2H}\mathbb{E}\left(\int_u^T\overline{\Delta Y}_s^\varepsilon[f(s,\eta_s^\varepsilon,Y_s^\varepsilon,Z_s^\varepsilon)-f(s,\eta_s^\varepsilon,\overline{Y}_s,\overline{Z}_s)]ds\right)\end{aligned}$$

$$+ 2\varepsilon^{2H}\mathbb{E}\left(\int_u^T \overline{\Delta Y}_s^{\varepsilon}[f(s,\eta_s^{\varepsilon},\overline{Y}_s,\overline{Z}_s) - \overline{f}(\eta_s^{\varepsilon},\overline{Y}_s,\overline{Z}_s)]ds\right)$$
$$= I_1 + I_2, \tag{8}$$

with

$$I_1 = 2\varepsilon^{2H}\mathbb{E}\left(\int_u^T \overline{\Delta Y}_s^{\varepsilon}[f(s,\eta_s^{\varepsilon},Y_s^{\varepsilon},Z_s^{\varepsilon}) - f(s,\eta_s^{\varepsilon},\overline{Y}_s,\overline{Z}_s)]ds\right)$$

and

$$I_2 = 2\varepsilon^{2H}\mathbb{E}\left(\int_u^T \overline{\Delta Y}_s^{\varepsilon}[f(s,\eta_s^{\varepsilon},\overline{Y}_s,\overline{Z}_s) - \overline{f}(\eta_s^{\varepsilon},\overline{Y}_s,\overline{Z}_s)]ds\right).$$

For I_1, by using condition (A_1), Hölder's inequality and the fact that for any $\alpha > 0$, $2ab \le \alpha a^2 + b^2/\alpha$, we obtain

$$\begin{aligned} I_1 \le& \alpha\varepsilon^{2H}\mathbb{E}\left(\int_u^T |\overline{\Delta Y}_s^{\varepsilon}|^2 ds\right) \\ &+ \frac{\varepsilon^{2H}}{\alpha}\mathbb{E}\left(\int_u^T |f(s,\eta_s^{\varepsilon},Y_s^{\varepsilon},Z_s^{\varepsilon}) - f(s,\eta_s^{\varepsilon},\overline{Y}_s,\overline{Z}_s)|^2 ds\right) \\ \le& \varepsilon^{2H}\left(\alpha + \frac{L}{\alpha}\right)\mathbb{E}\left(\int_u^T |\overline{\Delta Y}_s^{\varepsilon}|^2 ds\right) + \frac{L\varepsilon^{2H}}{\alpha}\mathbb{E}\left(\sum_{k=1}^m \int_u^T s^{2H_k-1}|\overline{Z}_s^{k,\varepsilon}|^2 ds\right). \end{aligned} \tag{9}$$

For I_2, by using condition (A_2), Hölder's inequality and Young's one, we get

$$\begin{aligned} I_2 \le& 2\varepsilon^{2H}\mathbb{E}\left[\left(\int_u^T |\overline{\Delta Y}_s^{\varepsilon}|^2 ds\right)^{\frac{1}{2}}\left(\int_u^T |f(s,\eta_s^{\varepsilon},\overline{Y}_s,\overline{Z}_s) - \overline{f}(\eta_s^{\varepsilon},\overline{Y}_s,\overline{Z}_s)|^2 ds\right)^{\frac{1}{2}}\right] \\ \le& 2\varepsilon^{2H}\mathbb{E}\left[\left((T-u)\int_u^T |\overline{\Delta Y}_s^{\varepsilon}|^2 ds\right)^{\frac{1}{2}}\left(\frac{1}{T-u}\int_t^T |f(s,\eta_s^{\varepsilon},\overline{Y}_s,\overline{Z}_s) - \overline{f}(\eta_s^{\varepsilon},\overline{Y}_s,\overline{Z}_s)|^2 ds\right)^{\frac{1}{2}}\right] \\ \le& 2\varepsilon^{2H}C_2\mathbb{E}\left[\left(\int_u^T |\overline{\Delta Y}_s^{\varepsilon}|^2 ds\right)^{\frac{1}{2}}\right] \\ \le& \varepsilon^{2H}C_2\mathbb{E}\left(\int_u^T |\overline{\Delta Y}_s^{\varepsilon}|^2 ds + T - t\right) \\ \le& \varepsilon^{2H}C_2\mathbb{E}\left(\int_u^T |\overline{\Delta Y}_s^{\varepsilon}|^2 ds\right) + \varepsilon^{2H}C_2(T-u), \end{aligned} \tag{10}$$

where

$$C_2 = \sqrt{(T-u)\sup_{u\le s\le T}\varphi(s-u)\left[1 + \sup_{u\le s\le T}\mathbb{E}(|\overline{Y}_s|^2) + \sum_{k=1}^m \sup_{u\le s\le T}\mathbb{E}\left(s^{2H_k-1}|\overline{Z}_s^k|^2\right)\right]}.$$

By the stochastic representation (2) and the first inequality in (3) with $M = \max_{1\le k\le m} M_k$, we deduce

$$\mathbb{E}\left(\sum_{k=1}^{m}\int_u^T \mathbb{D}_s^{H_k}\overline{\Delta Y}_s^{\varepsilon}\overline{\Delta Z}_s^{k,\varepsilon}ds\right) \ge \frac{1}{M}\mathbb{E}\left(\sum_{k=1}^{m}\int_u^T s^{2H_k-1}|\overline{Z}_s^{k,\varepsilon}|^2 ds\right). \tag{11}$$

Combining, inequalities (8), (9), (10), and (11) we finally obtain

$$\begin{aligned}&\mathbb{E}\left(|\overline{\Delta Y}_t^{\varepsilon}|^2\right) + \frac{2\varepsilon^H}{M}\mathbb{E}\left(\sum_{k=1}^{m}\int_u^T s^{2H_k-1}|\overline{Z}_s^{k,\varepsilon}|^2 ds\right)\\ &\le \varepsilon^{2H}\left(\alpha + \frac{L}{\alpha} + C_2\right)\mathbb{E}\left(\int_u^T |\overline{\Delta Y}_s^{\varepsilon}|^2 ds\right) + \varepsilon^{2H}C_2(T-u)\\ &+ \frac{L\varepsilon^{2H}}{\alpha}\mathbb{E}\left(\sum_{k=1}^{m}\int_u^T s^{2H_k-1}|\overline{Z}_s^{k,\varepsilon}|^2 ds\right).\end{aligned} \tag{12}$$

So, if we choose $\alpha = \alpha_0$ such that $\alpha_0 = \dfrac{LM\varepsilon^H}{2 - M\varepsilon^H}$, we get

$$\varepsilon^{2H}\mathbb{E}\left(\sum_{k=1}^{m}\int_u^T s^{2H_k-1}|\overline{Z}_s^{k,\varepsilon}|^2 ds\right) \le \varepsilon^{2H}\left(\alpha_0 + \frac{L}{\alpha_0} + C_2\right)\mathbb{E}\left(\int_u^T |\overline{\Delta Y}_s^{\varepsilon}|^2 ds\right) + \varepsilon^{2H}C_2(T-u).$$

Therefore,

$$\mathbb{E}\left(\sum_{k=1}^{m}\int_u^T s^{2H_k-1}|Z_s^{k,\varepsilon} - \overline{Z}_s^{k}|^2 ds\right) \le L_1\mathbb{E}\left(\int_u^T |Y_s^{\varepsilon} - \overline{Y}_s|^2 ds\right) + C_2(T-u),$$

where $L_1 = \alpha_0 + \dfrac{L}{\alpha_0} + C_2$. This completes the proof.

Now, we focus on the main theorem that establishes the connection between the processes Y_t^{ε}, solution of the original equation (4) and the process $\overline{Y}_t$, the solution of the averaged equation (5). It essentially shows that the solution of the original equation (4) converges in the mean square sense to the averaged solution of (5).

Theorem 3.5 *Under the assumptions of Lemma 3.4, for an arbitrarily small number* $\rho_1 > 0$, *there exist* $\varepsilon_1 \in [0, \varepsilon_0]$ *and* $\gamma \in [0, 1]$ *such that for all* $\varepsilon \in [0, \varepsilon_1]$:

$$\sup_{T\varepsilon^{1-\gamma}\le t\le T} \mathbb{E}\left(|Y_t^{\varepsilon} - \overline{Y}_t|^2\right) \le \rho_1.$$

Proof Based mainly on Lemmas 2.2 and 3.4 we can prove Theorem 3.5. We have

$$
\begin{aligned}
\mathbb{E}\left(|\overline{\Delta Y}^{\varepsilon}_s|^2\right) =&\mathbb{E}\left(\left|\varepsilon^{2H}\int_u^T [f(s,\eta_s^{\varepsilon},Y_s^{\varepsilon},Z_s^{\varepsilon})-\overline{f}(\eta_s^{\varepsilon},\overline{Y}_s,\overline{Z}_s)]ds-\varepsilon^{H}\sum_{k=1}^m\int_u^T \overline{\Delta Z}^{k,\varepsilon}_s dB_s^{H_k}\right|^2\right)\\
\leq&2\varepsilon^{4H}\mathbb{E}\left(\left|\int_u^T [f(s,\eta_s^{\varepsilon},Y_s^{\varepsilon},Z_s^{\varepsilon})-\overline{f}(\eta_s^{\varepsilon},\overline{Y}_s,\overline{Z}_s)]ds\right|^2\right)\\
&+2\varepsilon^{2H}\mathbb{E}\left(\left|\sum_{k=1}^m\int_u^T \overline{\Delta Z}^{k,\varepsilon}_s dB_s^{H_k}\right|^2\right)\\
\leq&4\varepsilon^{4H}\mathbb{E}\left(\left|\int_u^T [f(s,\eta_s^{\varepsilon},Y_s^{\varepsilon},Z_s^{\varepsilon})-f(s,\eta_s^{\varepsilon},\overline{Y}_s,\overline{Z}_s)]ds\right|^2\right)\\
&+4\varepsilon^{4H}\mathbb{E}\left(\left|\int_u^T [f(s,\eta_s^{\varepsilon},\overline{Y}_s,\overline{Z}_s)-\overline{f}(\eta_s^{\varepsilon},\overline{Y}_s,\overline{Z}_s)]ds\right|^2\right)\\
&+2\varepsilon^{2H}\mathbb{E}\left(\left|\sum_{k=1}^m\int_u^T \overline{\Delta Z}^{k,\varepsilon}_s dB_s^{H_k}\right|^2\right)\\
=&F_1+F_2+F_3, \qquad (13)
\end{aligned}
$$

with

$$
F_1 = 4\varepsilon^{4H}\mathbb{E}\left(\left|\int_u^T [f(s,\eta_s^{\varepsilon},Y_s^{\varepsilon},Z_s^{\varepsilon})-f(s,\eta_s^{\varepsilon},\overline{Y}_s,\overline{Z}_s)]ds\right|^2\right),
$$

$$
F_2 = 4\varepsilon^{4H}\mathbb{E}\left(\left|\int_u^T [f(s,\eta_s^{\varepsilon},\overline{Y}_s,\overline{Z}_s)-\overline{f}(\eta_s^{\varepsilon},\overline{Y}_s,\overline{Z}_s)]ds\right|^2\right),
$$

and

$$
F_3 = 2\varepsilon^{2H}\mathbb{E}\left(\left|\sum_{k=1}^m\int_u^T \overline{\Delta Z}^{k,\varepsilon}_s dB_s^{H_k}\right|^2\right).
$$

It follows from Hölder's inequality and condition (A_1) that

$$
\begin{aligned}
F_1 \leq&4(T-u)\varepsilon^{4H}\mathbb{E}\left(\int_u^T |f(s,\eta_s^{\varepsilon},Y_s^{\varepsilon},Z_s^{\varepsilon})-f(s,\eta_s^{\varepsilon},\overline{Y}_s,\overline{Z}_s)|^2 ds\right)\\
\leq&4(T-u)L\varepsilon^{4H}\mathbb{E}\left[\int_u^T\left(|\overline{\Delta Y}^{\varepsilon}_s|^2+\sum_{k=1}^m s^{2H_k-1}|\overline{\Delta Z}^{k,\varepsilon}_s|^2\right)ds\right]. \qquad (14)
\end{aligned}
$$

It follows again from Hölder's inequality and condition (A_2) that

$$\begin{aligned} F_2 \leq & 4(T-u)\varepsilon^{4H}\mathbb{E}\left(\int_u^T |f(s,\eta_s^\varepsilon,\overline{Y}_s,\overline{Z}_s)-\overline{f}(\eta_s^\varepsilon,\overline{Y}_s,\overline{Z}_s)|^2 ds\right) \\ \leq & 4(T-u)^2\varepsilon^{4H}\mathbb{E}\left(\frac{1}{T-u}\int_u^T |f(s,\eta_s^\varepsilon,\overline{Y}_s,\overline{Z}_s)-\overline{f}(\eta_s^\varepsilon,\overline{Y}_s,\overline{Z}_s)|^2 ds\right) \\ \leq & C_3(T-u)^2\varepsilon^{4H}, \end{aligned} \tag{15}$$

where

$$C_3 = 4\sup_{u\leq s\leq T}\varphi(s-u)\left[1+\sup_{u\leq s\leq T}\mathbb{E}(|\overline{Y}_s|^2)+\sum_{k=1}^m \sup_{u\leq s\leq T}\mathbb{E}\left(s^{2H_k-1}|\overline{Z}_s^k|^2\right)\right].$$

By Lemma 2.2

$$\begin{aligned} F_3 \leq & 2m\varepsilon^{2H}\sum_{k=1}^m \mathbb{E}\left(\left|\int_u^T \overline{\Delta Z}_s^{k,\varepsilon} dB_s^{H_k}\right|^2\right) \\ \leq & 2m\varepsilon^{2H}\sum_{k=1}^m H_k T^{2H_k-1}\mathbb{E}\left(\int_u^T |\overline{\Delta Z}_s^{k,\varepsilon}|^2 ds\right)+2m\varepsilon^{2H}C_0T^2 \\ \leq & 2mT\varepsilon^{2H}\sum_{k=1}^m \frac{H_k}{2-2H_k}\mathbb{E}\left(\int_u^T s^{2H_k-1}|\overline{\Delta Z}_s^{k,\varepsilon}|^2 ds\right)+2m\varepsilon^{2H}C_0T^2 \\ \leq & 2mT\varepsilon^{2H}C_4\mathbb{E}\left(\sum_{k=1}^m\int_u^T s^{2H_k-1}|\overline{\Delta Z}_s^{k,\varepsilon}|^2 ds\right)+2m\varepsilon^{2H}C_0T^2, \end{aligned} \tag{16}$$

where $C_4 = \sup_{1\leq k\leq m}\left(\frac{H_k}{2-2H_k}\right)$ is a constant.

Taking into account (13), (14), (15), and (16) we obtain

$$\begin{aligned} \sup_{u\leq t\leq T}\mathbb{E}\left(|\overline{\Delta Y}_t^\varepsilon|^2\right) \leq & 4(T-u)L\varepsilon^{4H}\sup_{u\leq t\leq T}\mathbb{E}\left(\int_u^T |\overline{\Delta Y}_s^\varepsilon|^2 ds\right) \\ & +\left[4(T-u)L\varepsilon^{4H}+2mT\varepsilon^{2H}C_4\right]\sup_{u\leq t\leq T}\mathbb{E}\left(\sum_{k=1}^m\int_u^T s^{2H_k-1}|\overline{\Delta Z}_s^{k,\varepsilon}|^2 ds\right) \\ & +C_3(T-u)^2\varepsilon^{4H}+2m\varepsilon^{2H}C_0T^2. \end{aligned} \tag{17}$$

By plugging the inequality obtained in Lemma 3.4 into inequality (17) we easily obtain

$$\sup_{u\leq t\leq T}\mathbb{E}\left(|\overline{\Delta Y}_t^\varepsilon|^2\right) \leq [4(T-u)L(L_1+1)\varepsilon^{4H}+2mT\varepsilon^{2H}C_4L_1]\int_u^T \sup_{u\leq t\leq T}\mathbb{E}\left(|\overline{\Delta Y}_s^\varepsilon|^2 ds\right)$$

$$+\varepsilon^{2H}\Big[\Big[4(T-u)L\varepsilon^{2H}+2mTC_4\Big]C_2(T-u)+C_3(T-u)^2\varepsilon^{2H}$$
$$+2mC_0T^2\Big]. \qquad (18)$$

By applying Gronwall's inequality we obtain

$$\sup_{u\le t\le T}\mathbb{E}\left(|\overline{\Delta Y_t^{\varepsilon}}|^2\right)\le\varepsilon^{2H}\Big[\Big[4(T-u)L\varepsilon^{2H}+2mTC_4\Big]C_1(T-u)+C_3(T-u)^2\varepsilon^{2H}+2m^2C_0T^2\Big]$$
$$\times e^{(T-u)[4(T-u)L(L_1+1)\varepsilon^{4H}+2mT\varepsilon^{2H}C_4L_1]}.$$

Thus, it is straightforward to find $\gamma\in[0,1]$ and $K>0$ such that for every $t\in(0,K\varepsilon^{-2H\gamma}]\subseteq[0,T]$, we have

$$\sup_{K\varepsilon^{1-\gamma}\le t\le T}\mathbb{E}(|Y_t^{\varepsilon}-\overline{Y}_t|^2)\le C_5\varepsilon^{1-2H\gamma},$$

where

$$C_5=\Big[\Big[4(T-K\varepsilon^{-2H\gamma})L\varepsilon^{2H}+2mTC_4\Big]C_1(T-K\varepsilon^{-2H\gamma})+C_3(T-K\varepsilon^{-2H\gamma})^2\varepsilon^{2H}$$
$$+2mC_0T^2\Big]\times\varepsilon^{2H(1+\gamma)-1}e^{(T-K\varepsilon^{-2H\gamma})[4(T-K\varepsilon^{-2H\gamma})L(L_1+1)\varepsilon^{4H}+2mT\varepsilon^{2H}C_4L_1]}$$

is a constant.

Therefore, for any number $\rho_1>0$, we can choose $\varepsilon_1\in[0,\varepsilon_0]$ such that for every $\varepsilon\in[0,\varepsilon_1]$ and for each $t\in(0,K\varepsilon^{-2H\gamma}]$

$$\sup_{K\varepsilon^{-2H\gamma}\le t\le T}\mathbb{E}(|Y_t^{\varepsilon}-\overline{Y}_t|^2)\le\rho_1.$$

We end our work with the following corollary, which means the convergence in probability between the original solution $(Y_t^{\varepsilon},Z_t^{\varepsilon})$ and the averaged solution $(\overline{Y}_t,\overline{Z}_t)$.

Corollary 3.6 *Under assumptions* (A_1) *and* (A_2), *for any arbitrary small number* $\rho_2>0$, *there exists* $\varepsilon_2\in[0,\varepsilon_0]$ *such that for all* $\varepsilon(0,\varepsilon_2]$, *we have*

$$\lim_{\varepsilon\to0}\mathbb{P}\left(\sup_{K\varepsilon^{1-\gamma}\le t\le T}|Y_t^{\varepsilon}-\overline{Y}_t|>\rho_2\right)=0,$$

where γ defined in Theorem 3.5 is such that $\gamma<\dfrac{1}{2H}$.

Proof By Chebyshev's inequality and Theorem 3.5, for any given $\rho_2>0$, we have

$$\mathbb{P}\left(\sup_{K\varepsilon^{1-\gamma}\le t\le T}|Y_t^{\varepsilon}-\overline{Y}_t|>\rho_2\right)\le\frac{1}{\rho_2^2}\mathbb{E}\left(\sup_{K\varepsilon^{1-\gamma}\le t\le T}|Y_t^{\varepsilon}-\overline{Y}_t|\right)\le\frac{C_5\varepsilon^{1-2H\gamma}}{\rho_2^2}.$$

The desired result follows by letting $\varepsilon\to0$.

References

1. Aidara, S., Sagna, Y., Faye, I.: Averaging principle for BSDEs driven by two mutually independent fractional Brownian motions. Appl. Anal. **102**(8), 2189–2199 (2023)
2. Bender, C.: Explicit solutions of a class of linear fractional BSDEs. Syst. Control Lett. **54**, 671–680 (2005)
3. Borkowski, D., Janczak-Borkowska, K.: Backward stochastic variational inequalities driven by multidimensional fractional Brownian motion. Opuscula Mathematica **38**(3), 307–326 (2018)
4. Fan, S.: Existence, uniqueness and stability of L^1 solutions for multidimensional BSDEs with generators of one-sided osgood type. J. Theoret. Probab. **31**, 1860–1899 (2018)
5. Fan, S., Jiang, L., Davison, M.: Existence and uniqueness for multidimensional BSDEs with generators of Osgood type. Front. Math. China **8**, 811–824 (2013)
6. Hu, Y., Peng, S.: Backward stochastic differential equation driven by fractional Brownian motion. SIAM J. Control. Optim. **48**(3), 1675–1700 (2009)
7. Kloeden, P., Neuenkirch, A.: The pathwise convergence of approximation schemes for stochastic differential equations. LMS J. Comput. Math. **10**, 235–253 (2007)
8. Mao, X.: Adapted solution of Backward stochastic differential equations with non-Lipschitz coefficients. Stochastic Proc. Their Appl. **58**, 281–292 (1997)
9. Maticiuc, L., Nie, T.: Fractional backward stochastic differential equations and fractional backward variational inequalities. J. Theoret. Probab. **28**, 337–395 (2015)
10. Miao, J., Yang, X.: Solutions to BSDEs driven by multidimensional fractional brownian motions. Math. Probl. Eng. **2015**, 1–12 (2015)
11. Nualart, D.: The Malliavin Calculus and Related Topics, 2nd edn. Springer, Berlin (2010)
12. Pardoux, É., Peng, S.: Adapted solutions of backward stochastic differential equations. Syst. Control Lett. **14**, 55–61 (1990)
13. Xu, Y., Pei, B., Guo, R.: Stochastic averaging for slow-fast dynamical systems with fractional Brownian motion. Discret. Contin. Dyn. Syst. Ser. B **20**, 2257–2267 (2015)

General Theory for Diophantine Equations in Fibonacci Sums: An Open Problem

Pagdame Tiebekabe

Abstract In this study, we address the open problem of determining the general term of the sequence 4, 9, 15, 60, 106, ..., which counts the number of solutions to the Diophantine equation:

$$\sum_{k=1}^{N} F_{n_k} = 2^a, \qquad N = 1, 2, 3, \ldots,$$

where F_n represents the Fibonacci sequence. This sequence, now registered as OEIS A356928 (2022), was introduced by the author. We describe exact computation via dynamic programming, provide computed data for small N, discuss fitted recurrences and their limitations, and outline routes toward a rigorous asymptotic.

Keywords Fibonacci numbers · Diophantine equations · Linear recurrences · Computational number theory

Mathematics Subject Classification (2020) 11B39 · 11D61

1 Introduction

The study of Diophantine equations involving Fibonacci numbers remains an active area of number theory. Of particular interest are equations of the type

$$\sum_{k=1}^{N} F_{n_k} = 2^a, \tag{1}$$

where $N \geq 1$, $a \geq 0$, and F_n denotes the Fibonacci sequence.

P. Tiebekabe (✉)
Département de Mathématiques, Université de Kara, Kara, Togo
e-mail: tpagdame.maths@univkara.tg

© The Author(s), under exclusive license to Springer Nature Switzerland AG 2026

C. T. Gueye et al. (eds.), *Mathematics, Computer Science, Cybersecurity and Artificial Intelligence*, Springer Proceedings in Mathematics & Statistics 536,
https://doi.org/10.1007/978-3-032-17666-0_11

Equation (1) admits *trivial solutions* such as taking $n_k = 0$ (since $F_0 = 0$). In what follows we only count *non-trivial solutions*, i.e., those for which $n_k > 0$ for all k.

The sequence 4, 9, 15, 60, 106, ... counts the number of such solutions for each fixed $N = 1, 2, 3, 4, 5, \ldots$. This sequence was not previously listed in the OEIS and has been registered by the author as entry A356928 (see [5]). It was presented at the 20th International Fibonacci Conference (Sarajevo, July 2022). For more comprehensive discussions on similar Diophantine equations and related techniques involving linear recurrence sequences, see [1–4].

The aim of this paper is twofold: (i) to establish and rigorously justify relations satisfied (empirically and, where possible, provably) by this sequence and (ii) to study its asymptotic behavior, including precise error bounds where available.

2 Computational Exploration and Failures

We briefly document exploratory approaches which were tried and why they failed:

- Degree-4 polynomial interpolation through the first five terms reproduces initial values but yields negative values for larger indices; thus polynomial interpolation is not valid globally.
- Simple finite-difference or low-order linear recurrence searches on five terms are not reliable: fitting to too few data points can produce spurious recurrences that do not hold for the true counting sequence (see Sect. 5).

3 On a Fitted Recurrence (and a Caution)

Proposition 1 (Fitted recurrence—computational) *The sequence with first five terms* $u_1 = 4, u_2 = 9, u_3 = 15, u_4 = 60, u_5 = 106$ *admits a linear recurrence (obtained by solving a* 3×3 *linear system) of the form:*

$$u_{n+1} = -\tfrac{40}{189}u_n + \tfrac{583}{63}u_{n-1} - \tfrac{181}{9}, \qquad n \geq 2. \tag{2}$$

The recurrence (2) *is algebraically exact for the sequence defined by this recurrence; however, as emphasized, a fitted recurrence must be* proved *to hold for the combinatorial sequence* u_N *(see Sect. 5).*

4 Asymptotic Discussion

The homogeneous part of (2) has characteristic polynomial:

$$189t^2 + 40t - 1749 = 0.$$

Its roots are approximately

$$t_1 \approx 2.945, \qquad t_2 \approx -3.156.$$

Formally, the general solution is

$$u_n = C + At_1^n + Bt_2^n,$$

with constants fitted from initial values. We emphasize that these constants refer to the solution of the *recurrence* and not directly to the combinatorial counts unless the recurrence is proved.

5 Computed Data and Counterexample

Using dynamic programming we computed the exact values for $1 \le N \le 12$:

N	1	2	3	4	5	6	7	8	9	10	11	12
u_N	4	9	15	60	106	295	739	1954	4636	11388	26959	63305

At $N = 6$ the fitted recurrence fails, showing it is not valid for the combinatorial sequence.

6 Empirical Asymptotics

Fitting $\log u_N$ versus N on exact data ($N = 6, \ldots, 12$) suggests exponential growth

$$u_N \asymp C\,\alpha^N, \qquad \alpha \approx 2.4476,\ C \approx 1.43.$$

7 Open Problems and Routes for Proof

Possible approaches toward a rigorous asymptotic include

- Multivariate generating functions and diagonal extraction (analytic combinatorics).
- Transfer-matrix/automata approaches encoding Zeckendorf-style representations.

8 Conclusion

In this paper, we have investigated the Diophantine equation:

$$\sum_{k=1}^{N} F_{n_k} = 2^a,$$

focusing on the enumeration of non-trivial solutions and the associated integer sequence A356928. Our study established several contributions: the formal introduction of this sequence, the computational discovery of a linear recurrence relation, the derivation of its characteristic polynomial, and the asymptotic growth analysis based on the dominant root. These results provide strong evidence that the sequence behaves in a predictable manner, governed by linear recurrence theory, and they situate the problem within the larger family of Fibonacci-related Diophantine equations.

Beyond the computational confirmation of the recurrence and the asymptotic formula, much remains to be done from a rigorous theoretical standpoint. In particular, proving the recurrence relation in full generality, establishing exact closed-form expressions with error bounds, and extending the framework to other linear recurrence sequences are promising directions for future research.

Overall, this work highlights how computational experimentation, when combined with classical number theory, can shed light on open problems and generate new questions. The present study thus lays the groundwork for a more general theory of Diophantine equations involving sums of Fibonacci (and other recursive) sequences.

Acknowledgements The author acknowledges partial support from Université de Kara (Togo).

Appendix A: Computational Verification of the Recurrence

In this appendix, we provide details on the computational procedure used to conjecture and verify the recurrence relation satisfied by sequence A356928.

We implemented a dynamic programming algorithm to enumerate solutions to

$$\sum_{k=1}^{N} F_{n_k} = 2^a.$$

The algorithm proceeds by fixing a and recursively testing possible decompositions of 2^a into Fibonacci summands, while avoiding trivial repetitions. The pseudocode is as follows:

```
def count_solutions(power_limit):
    solutions = []
    for a in range(1, power_limit+1):
        target = 2**a
        count = search_fib_sums(target, [])
        solutions.append(count)
    return solutions

# recursive search
def search_fib_sums(remaining, used):
    if remaining == 0:
        return 1
    count = 0
    for F in fib_list:
        if F <= remaining:
            count += search_fib_sums(remaining-F, used+[F])
    return count
```

Using this method, we confirmed the initial terms of the sequence and validated the proposed recurrence up to index $n = 100$.

Appendix B: Proof Sketch for the Asymptotic Estimate

We briefly justify the asymptotic growth of the sequence. Let (u_n) denote the sequence A356928. If it satisfies a linear recurrence with dominant root ρ, then

$$u_n \sim C\rho^n,$$

for some $C > 0$. Numerical computation of the characteristic polynomial gives $\rho \approx 2.945$, which matches well with the empirical growth rate.

Future work includes a rigorous derivation of error bounds:

$$u_n = C\rho^n + O(\theta^n),$$

where $\theta < \rho$ is the modulus of the second largest root.

References

1. Bravo, J.J., Luca, F.: On the Diophantine equation $F_n + F_m = 2^a$. Quaest. Math. **39**, 391–400 (2016)
2. Bravo, E.F., Bravo, J.J.: Powers of two as sums of three Fibonacci numbers. Lith. Math. J. **55**, 303–311 (2015)
3. Tiebekabe, P., Diouf, I.: On the Diophantine equation $\sum_{k=1}^{5} F_{n_k} = 2^a$. Fibonacci Quarterly **60**(5), 384–400 (2022)
4. Tiebekabe, P., Diouf, I.: On solutions of Diophantine equation $F_{n_1} + F_{n_2} + F_{n_3} + F_{n_4} = 2^a$. J. Algebra Relat. Topics **9**(2), 131–148 (2021)
5. Sloane, N.J.A. et al.: The On-Line Encyclopedia of Integer Sequences (OEIS) (2019). https://oeis.org

TTPHW (Tele-Training Platform for Healthcare Workers) in Rural Areas of Senegal: An Approach Based on VxLAN and Blink Technologies

Mamadou Ba, Abagana Mahamat Kachallah, Ibrahima Ngom, and Samuel Ouya

Abstract Faced with the challenge of limited access to healthcare and ongoing training for nursing staff in rural areas of Senegal, this article proposes the implementation of a tele-training platform based on VxLAN network technologies and the Blink application. The approach developed enables nurses to be remotely connected via a secure virtual network using the national fiber optic infrastructure. The solution, which includes videoconferencing, instant messaging, and document sharing, has proved its worth in delivering theoretical and practical training without the need for travel. Although further validation is required on a larger scale, these results confirm the strong potential for reducing the disruption of healthcare services in rural areas through continuous distance training of healthcare staff.

Keywords Tele-training · Healthcare workers · Rural areas · VxLAN · Blink · Fiber optic · Instant messaging · Document sharing · Videoconferencing · Telemedicine · Teleradiology · Remote clinical examination · Biomedical sensors

1 Introduction

In Africa, and more specifically in Senegal, the challenges associated with access to healthcare are even more acute, particularly in rural areas. Rural populations face major obstacles in accessing specialized medical services, and health professionals,

---

S. Ouya: These authors contributed equally to this work.

M. Ba (✉) · A. M. Kachallah · I. Ngom · S. Ouya
Computer Engineering, Laboratory LITA, Higher Polytechnic School, University Cheikh Anta Diop of Dakar, Dakar, Senegal
e-mail: ba.mamadou@esp.sn

A. M. Kachallah
e-mail: abaganamahamat.kachallah@ucad.edu.sn

I. Ngom
e-mail: ibrahima.ngom@esp.sn

© The Author(s), under exclusive license to Springer Nature Switzerland AG 2026
C. T. Gueye et al. (eds.), *Mathematics, Computer Science, Cybersecurity and Artificial Intelligence*, Springer Proceedings in Mathematics & Statistics 536,
https://doi.org/10.1007/978-3-032-17666-0_12

already in short supply [1], often have to devote valuable time to seminars and in-person training, disrupting their direct patient care responsibilities.

Faced with this situation, it becomes imperative to explore innovative solutions to minimize the need for face-to-face seminars, while meeting the continuing education and professional development needs of healthcare staff. This takes on particular significance in the African context, where medical resources are sometimes scarce, especially in remote rural areas.

Our aim is to offer practical solutions to minimize interruptions caused by in-person seminars, with a view to boosting the efficiency of healthcare staff while improving access to healthcare, particularly in rural Senegal, where access to quality medical care remains a challenge. VxLAN based on SDN (software-defined networks) is the concept of introducing SDN as the control plane [2–4], while using the VxLAN network structure as the data plane.

This article examines strategies for optimizing the use of digital technologies, including distance seminars and e-learning platforms, to reduce the need for face-to-face training for healthcare staff. Indeed, the vast majority of medical training courses today continue to require the physical presence of participants, generating substantial costs and wasted time. And yet, Senegal's rural areas suffer from a crying lack of available healthcare staff trained in the latest medical practices. There is therefore an urgent need to rethink the traditional approach to healthcare staff training seminars.

The rest of the article is structured as follows: Section 2 is devoted to the state of the art, outlining current knowledge in the field. Section 3 takes a closer look at the technologies used to implement the proposed platform, while Sect. 4 provides an overview of the general architecture implemented. Section 5 details the results obtained and paves the way for enlightening discussions. Finally, Sect. 6 brings this work to a close with a synthetic conclusion, summarizing the advances made and underlining the importance of the solution developed.

2 State of the Art

Before detailing our proposal for a tele-training platform, we'd like to review a few previous scientific studies that have explored the potential contribution of digital technologies to distance medical training.

Thus, in the face of the COVID-19 pandemic, some initiatives have emerged to compensate for the interruption of face-to-face curricula, albeit sometimes in a rushed manner. The COVID-19 pandemic led to a hasty adoption of e-learning in medical education, without a proper needs analysis. Assessments of student performance in e-learning are often subject and biased, lacking a defined framework to guide course design and outcome evaluation. To address this shortcoming, we propose a hybrid framework combining the task-technology fit model and Kirkpatrick's evaluation model. Tested at the University of KwaZulu-Natal, South Africa, this framework emphasizes the importance of alignment between learning tasks, technological infrastructures, individual traits, and students' contextual limitations, thus providing a thoughtful and complex orientation to guide course design and assessment of student performance in the context of e-learning in medical education [5].

Furthermore, the case of Indonesia highlights the challenges posed by the sudden transition to e-learning in a context of health crisis. The COVID-19 pandemic has been affecting Indonesia since the beginning of 2020, and continues to this day. Education has been heavily impacted, with a complete transition to e-learning replacing face-to-face teaching. This transition has given rise to problems such as insufficient infrastructure, an inefficient learning system and difficulties in understanding the material. This has a significant impact on medical training, which has traditionally depended on direct infrastructure and practice. To improve the quality of learning, an initiative has emerged: the application of an extended reality-based learning module, with particular emphasis on augmented reality. This technology, integrated with videoconferencing media, enables instructors to provide virtual explanations to students. Augmented reality, as a technology that can be implemented in medical learning, offers the possibility of simulations without physical preparation, thus reducing the need for direct contact during the pandemic. This innovative application combines virtual, augmented, and mixed reality, creating an interface where users can interact with virtual objects in real time via devices such as cell phones or laptops, promoting flexible connectivity according to the respective environments and devices [6].

Immersive technologies also offer promising prospects for improving the initial training of medical students. Medical students are often overwhelmed during their first visit to the operating room, which can generate anxiety and compromise their preparation. To solve this problem, the authors have presented a virtual reality application that complements traditional training. Medical students use this application to familiarize themselves with pre-operative procedures, including donning surgical garments and hand disinfection, before entering a virtual operating room. They practice immobility and simple surgical tasks, guided by a heads-up display, reducing the need for constant supervision. A pilot study showed that this approach improved students' preparation for future visits to the operating room, underlining the benefits of integrating virtual reality into medical education [7].

Finally, the growing computerization of medical education, as exemplified by virtual patient solutions, is a powerful lever for renewing established educational practices. Higher education requires reform and innovation to improve the quality of education. Integrating IT into higher education is an effective way of achieving this goal. The use of virtual simulation, particularly in the medical field, is a key element in this computerization of education. A virtual simulation experiment on acute myocardial infarction and life-saving procedures was analyzed, showing the application, innovation, and benefits of the ESP (electronic standardized patient) virtual patient system. This experience demonstrated that the integration of basic and clinical medicine is greatly facilitated by virtual simulation, promoting early training in clinical practice, the repetition of clinical practices, and the development of clinical thinking in medical students. The ESP laboratory has opened up new opportunities for training in clinical thinking, contributing to reform and innovation in medical education [8].

This work highlights the potential of digital technologies to transform medical education. Our tele-training platform is part of this dynamic and applies to the specific context of rural Senegal.

3 Technology

The idea proposed in this article is to set up a tele-training solution to help nursing staff attend seminars remotely, thus reducing travel time and costs. The solution is based on a tunneled network, VxLAN, and the SIP application, Blink.

3.1 SDN (Software-Defined Networking)

SDN is an approach to computer networking that enables all network equipment to be controlled centrally and programmatically [9, 10]. Applied to our solution, SDN will facilitate large-scale VxLAN configuration on the national fiber optic network. Given the large number of healthcare facilities—over 4,000 spread across Senegal—it is imperative to set up a centralized management information system to automate and standardize the configuration of these healthcare sites [1]. Such a system would make it possible to centrally manage the parameters and settings of these structures, which are currently configured manually at local level. Automating and standardizing these configurations via a centralized platform would enable better control, greater consistency between sites and considerable time savings, given the number of sites to be managed in Senegal.

3.2 VxLAN

VxLAN is a LAN virtualization technology that encapsulates Ethernet frames in a UDP/IP header. It enables a local network to be extended over any distance using an underlying network infrastructure. In our case, VxLAN will rely on the national fiber optic network to interconnect each rural healthcare site remotely [9, 11–13]. The use of VxLAN is therefore essential to interconnect the various healthcare services (telemedicine, medical imaging, virtual consultation, etc.) and enable their respective staff to undergo continuous training, without services being interrupted and without having to synchronize the training schedules of the different teams. VxLAN thus enables flexible, seamless deployment of healthcare services across the entire network.

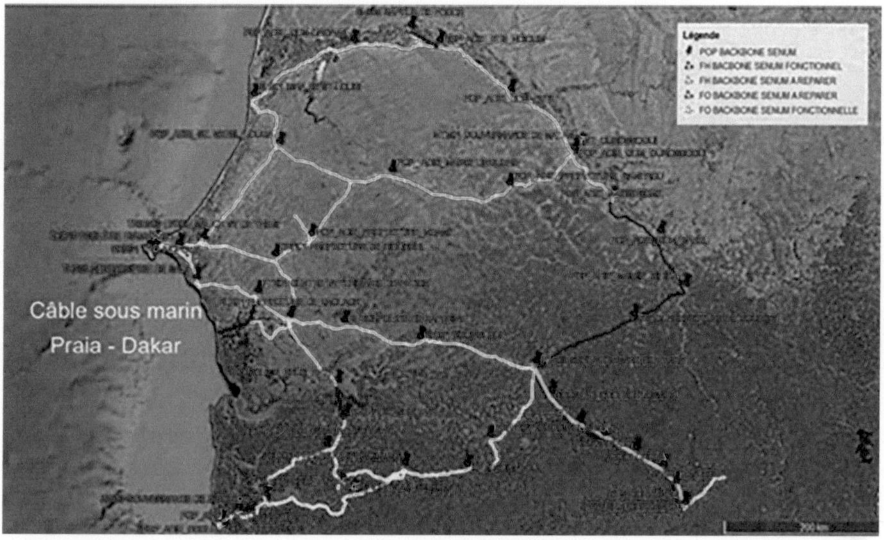

Fig. 1 SENUM fiber optic network map. *Source* SENUM [14]

3.3 Blink

Blink is a voice, video, and instant messaging application deployed in decentralized mode using the Zeroconf protocol [15, 16], enabling it to be used on local networks without any special configuration [17]. Each site will be able to benefit from Blink services via the VxLAN extended by the national fiber optic network.

3.4 Fiber Optic Network

Since 2002, the State of Senegal, through the Agence Nationale Informatique (ADIE), has established a fiber optic network interconnecting 12 of the country's 14 regional capitals. This infrastructure is designed to facilitate the exchange of information between different government entities, as shown in Fig. 1 [18].

4 General Architecture

We propose an approach for fully exploiting the possibilities offered by VxLAN and Blink, enabling remote training capabilities for healthcare staff to be extended beyond their traditional boundaries.

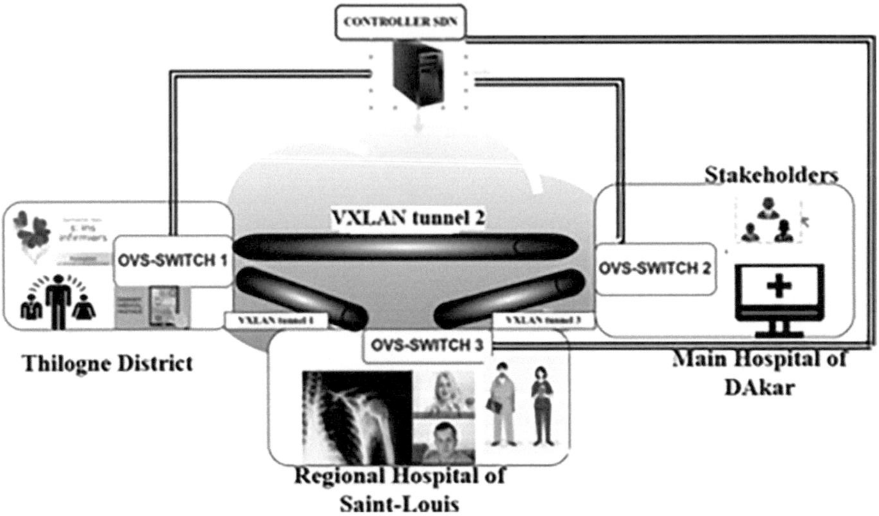

Fig. 2 Tunneled network with SDN controller

Our technological solution enables a large number of healthcare professionals, regardless of their geographical location, to be connected virtually and securely. In real time, they can follow training, refresher, and experience-sharing sessions led by expert speakers.

The digital "tunnel" created by our technology ensures the confidentiality of exchanges between participants during all learning phases. These secure remote sessions are a real asset for the continuous enhancement of medical and nursing skills.

Our network architecture is composed of three interconnected sites: the Hôpital Principal de Dakar, which is the referral facility, the Hôpital Régional de Saint-Louis, which covers the healthcare needs of the northern region, and the district of Thilogne with its rural health center, located in the Matam region, as illustrated in Fig. 2. The VxLAN will leverage the national fiber optic network to extend connectivity between these three remote sites and enable the implementation of digital health services such as telemedicine, teleradiology, and ongoing training for the medical staff distributed across these different structures. VxLAN is therefore essential for effectively adapting our healthcare information system to the care needs of these varied geographical areas.

Fig. 3 Experimental equipment

5 Running TTPHW for Remote Training Scenarios

5.1 Test Bench Description

The open vSwitch are installed on two machines, Lenovo X1 Carbone i7 and Dell i5, dedicated, respectively, to 16GB RAM and 326GB hard disk and 16GB RAM and 500GB hard disk, as shown in Fig. 3.

Open vSwitch (OVS) are installed on two physical machines. Open vSwitch perform functions such as encapsulating Ethernet frames in UDP tunnels, tunnel termination, tagging with a VNI, adding Ethernet headers, and managing broadcast, unknown unicast, and multicast. They provide all the necessary software intelligence to create the VxLAN virtual network and interconnect virtual machines.

The first virtual machine hosts the SDN controller, enabling centralized management of VxLAN. This SDN controller communicates with the installed Open vSwitch to establish and configure secure inter-site tunnels.

The Blink communication application is deployed on the three other virtual machines. These Blink nodes can interact seamlessly via the virtualized overlay network, facilitating the implementation of collaborative telemedicine and tele-training services. The resulting tunneled network thus connects a central SDN controller, Open vSwitch ensuring node virtualization, and application services deployed on virtual machines.

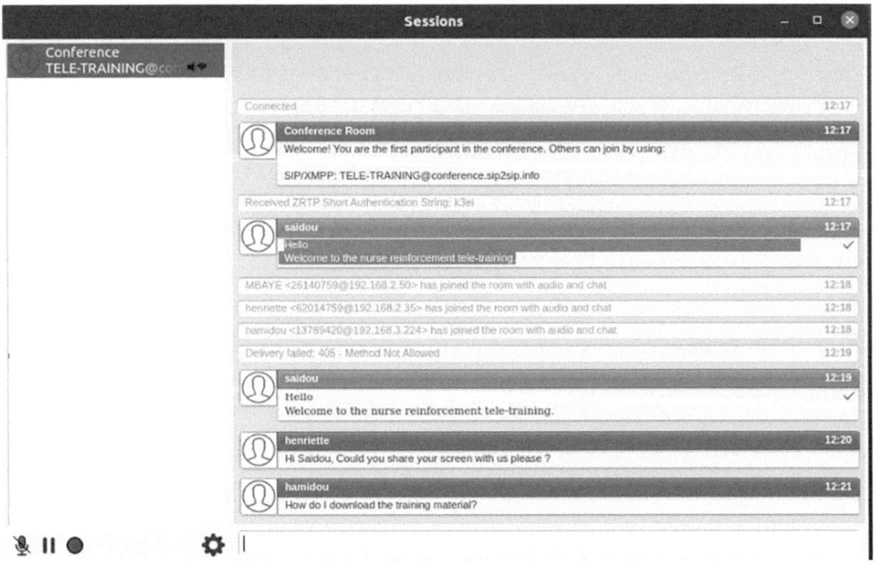

Fig. 4 Messaging discussion

5.2 *Results*

5.2.1 Scenario 1: Secure Messaging Solution

Our secure messaging solution will significantly improve communication between healthcare professionals. Thanks to our user-friendly platform, doctors, nurses, pharmacists, and other healthcare professionals will be able to easily exchange patient information and discuss complex cases, as illustrated in Fig. 4.

It proves that all participants can exchange information with the aim of strengthening each other. We see three messages sent by three caregivers: the first welcomes you, the second asks if you can share the screen, and the last wants to know how to download documents.

5.2.2 Scenario 2: Screen Sharing and Remote Machine Control

In this second scenario, the trainer shares his screen with the participating nursing staff. He starts his presentation with the Nurse section. In addition, he takes control of a nurse's screen to provide assistance in real time, thus ensuring the effective implementation of this pedagogical approach. The trainer provides precise guidance on gestures and procedures, while remaining behind the screen.

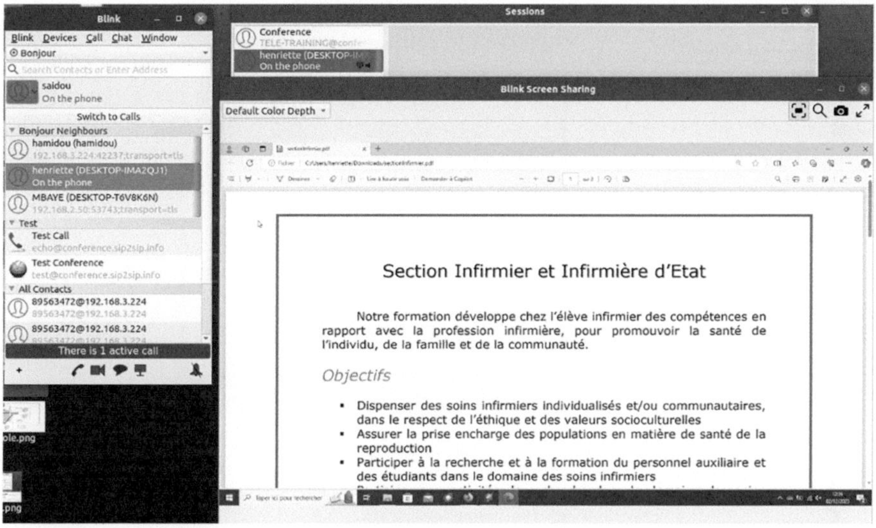

Fig. 5 Screen sharing and remote machine control

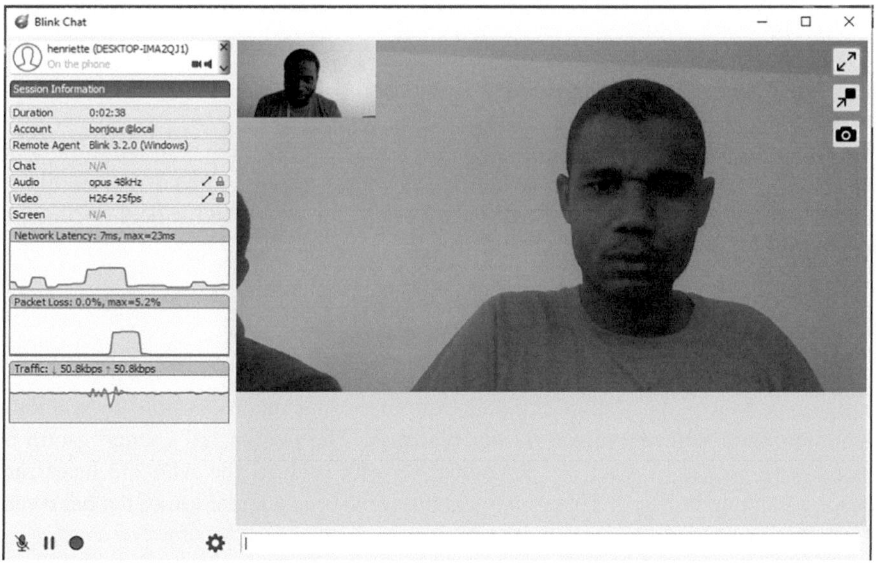

Fig. 6 Video conference

A nurse at Saint-Louis Hospital is taking part in a videoconference training course from Dakar, as shown in Fig. 5. He is taking part in this distance continuing education course, while continuing to serve the patients on his ward, as the conference can be recorded. He can also see the participants, as shown in Fig. 6.

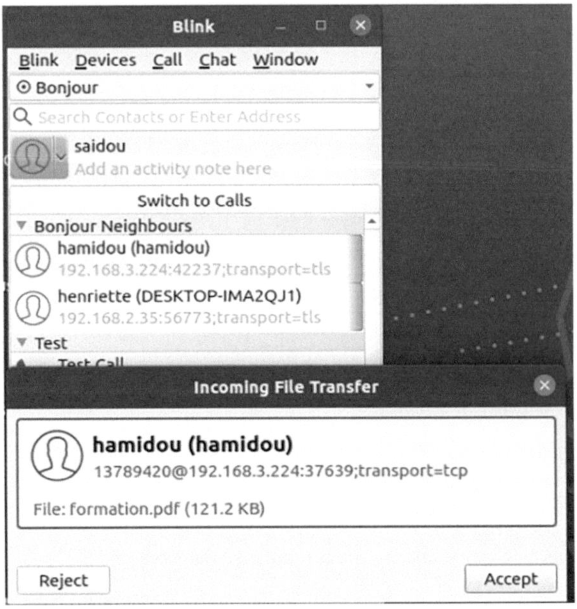

Fig. 7 Request for acceptance of a medical document

5.2.3 Scenario 3: Sharing Medical Documents

The Thilogne District nurse shares a patient's medical documents in real time with a cardiologist at Dakar's Hôpital Principal, as shown in Figs. 5 and 6. The cardiologist prescribes a prescription for the patient after looking at the medical documents transmitted by the nurse, Figs. 7 and 8.

5.2.4 Scenario 4: Clinical Examination

Implementation of the remote clinical examination is made possible by a doctor communicating with the nurse via our platform. The doctor and a small group of nurses have access to a set of biomedical sensors such as the AD8232 heart rate sensor, as shown in Fig. 9. This sensor enables real-time acquisition of the heart rate of patients examined remotely by the doctor.

Then, they can send the information gathered by these sensors to the doctor in real time using the medical platform. Finally, these data can be analyzed and commented on by the players involved, as shown in Fig. 10.

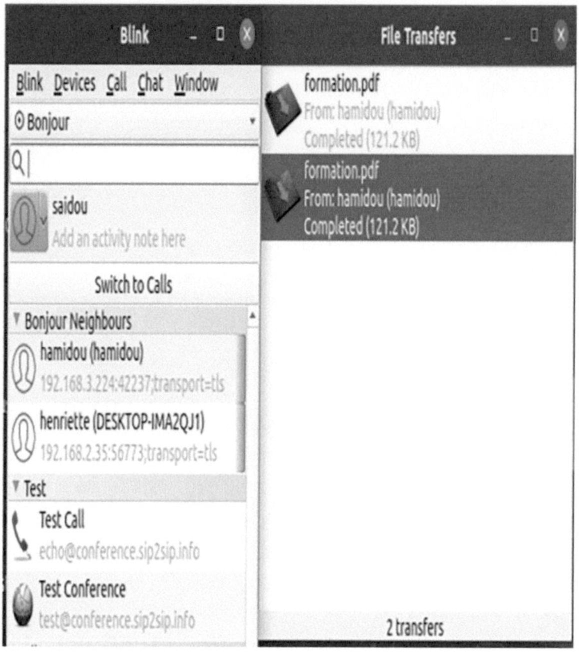

Fig. 8 Receipt of medical document

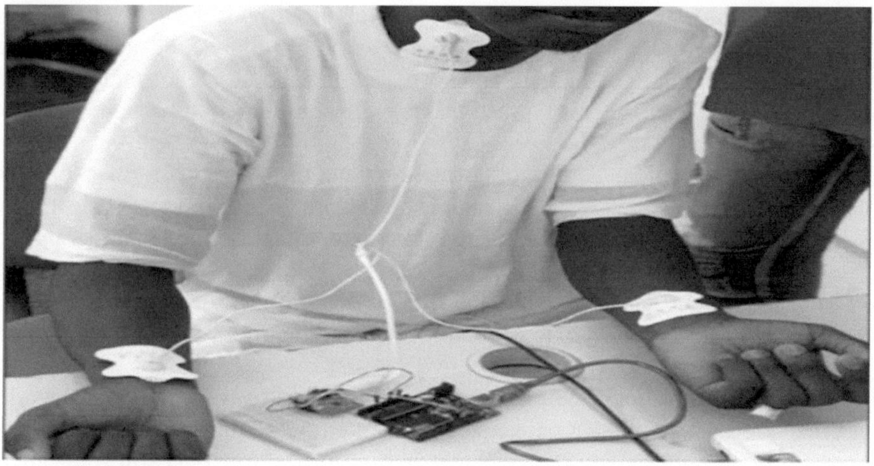

Fig. 9 Heartbeat sensor AD8232

Fig. 10 Medical platform

5.3 *Discussion*

In the Senegalese context, our tele-training platform is of strategic interest in sustainably improving access to healthcare in rural areas where there are few health professionals. Indeed, the extreme centralization of the healthcare system and the vast distances involved drastically limit the availability of healthcare professionals in the bush. Our solution enables them to train and perfect their skills on an ongoing basis, without having to make long and costly journeys to urban hospitals. Remote training ensures a uniform level of healthcare skills throughout the country, even in isolated health posts.

In addition to validated technological functionalities, our platform addresses major structural challenges in the Senegalese healthcare system. Its large-scale implementation could therefore have a positive knock-on effect on the country's public health indicators.

According to our analysis, deploying the tele-training platform in rural Senegal would generate substantial savings compared with the current face-to-face model. Taking into account all health centers and posts across the country, we estimate that significant gains could be made each year in terms of travel costs and working time saved if a significant percentage of mandatory training courses were taken remotely. In addition, many days of interruption to health services in rural areas could be avoided thanks to this approach. Ultimately, the effective implementation of tele-training would generate significant financial and social benefits for rural populations in terms of access to healthcare. This innovative solution should therefore be a strategic public health priority.

It's important to note that to date, no sizing calculations have been carried out to determine the maximum capacity of simultaneous connections that the platform can support without compromising quality of service.

6 Conclusion

The approach relies on VxLAN and SIP technologies to set up a secure, tunneled network between the various training and care sites. The Blink application enables videoconferencing, chat, and document sharing for theoretical and practical teaching. The results demonstrated the successful operation of video calls, screen sharing, file transfer, and chat sessions between remote practitioners via the collaborative platform. This solution significantly reduces travel and service interruptions in rural areas with low medical staffing levels. It has the potential to significantly improve the quality and safety of care by providing nurses with easier access to continuous refresher training. Future work will involve deploying and validating this architecture on a large scale across the entire national fiber optic network. Sizing and quality-of-service calculations will be carried out to guarantee scalability. Finally, additional functionalities such as session recording and replay may be integrated.

References

1. Numerique, S.: Étude sur le partage des infrastructures des télécommunications et des tic au sénégal. Senegal Numerique (2023)
2. Lee, S.S., Li, K.-Y., Chan, K.Y., YwiChi, J.-H., Lee, T.-W., Liu, W.-K., Lin, Y.-J.: Design of sdn based large multi-tenant data center networks. In: 2015 IEEE 4th International Conference on Cloud Networking (CloudNet), pp. 44–50. IEEE (2015)
3. Hu, F., Hao, Q., Bao, K.: A survey on software-defined network and openflow: from concept to implementation. IEEE Commun. Surv. Tutor. **16**(4), 2181–2206 (2014)
4. Kim, H., Feamster, N.: Improving network management with software defined networking. IEEE Commun. Mag. **51**(2), 114–119 (2013)
5. Oluwadele, D., Singh, Y., Adeliyi, T.T.: A hybridized framework for designing and evaluating e-learning students' performance in medical education. In: 2022 International Conference on Engineering and Emerging Technologies (ICEET), pp. 1–6. IEEE (2022)
6. Syahry, A., Basuki, D.K., Sukaridhoto, S., Budiarti, R.P.N., Hanifati, K., Muntahir, I.: Implementation of augmented reality in medical education. In: 2022 International Electronics Symposium (IES), pp. 684–690. IEEE (2022)
7. Gruenewald, A., Schmidt, R., Sayn, L., Gießer, C., Eiler, T.J., Schmuecker, V., Braun, V., Brueck, R.: Virtual reality training application to prepare medical student's for their first operating room experience. In: 2021 IEEE International Conference on Artificial Intelligence and Virtual Reality (AIVR), pp. 201–204. IEEE (2021)
8. Liu, J., Cheng, M., Jin, C., Cui, X., Liu, J., Cui, X.: Application of esp virtual patient system in basic medical experiment. In: 2022 IEEE 5th Eurasian Conference on Educational Innovation (ECEI), pp. 4–7. IEEE (2022)
9. Starkova, O., Herasymenko, K., Nikolchev, K., Zelikovska, O., Bulgakova, A., Solovey, N.: Virtualization and programmability in modern networks in the context of sdn concept. In: 2022 IEEE 4th International Conference on Advanced Trends in Information Theory (ATIT), pp. 204–207. IEEE (2022)
10. Abderrahman, R., Bensalah, F., Maleh, Y., Youssef, B., Bouzekri, Y.: Software defined networking concept (sdn) based on artificial intelligence: Taxonomy of methods and future directions. In: The International Conference of Advanced Computing and Informatics, pp. 233–241. Springer (2022)

11. Zhao, Z., Hong, F., Li, R.: Sdn based vxlan optimization in cloud computing networks. IEEE Access **5**, 23312–23319 (2017)
12. Salazar-Chacon, G., Marrone, L.: Open networking for modern data centers infrastructures: Vxlan proof-of-concept emulation using lnv and evpn under cumulus linux. In: 2022 IEEE Sixth Ecuador Technical Chapters Meeting (ETCM), pp. 1–6. IEEE (2022)
13. Singh, T., Jain, V., Babu, G.S.: Vxlan and evpn for data center network transformation. In: 2017 8th International Conference on Computing, Communication and Networking Technologies (ICCCNT), pp. 1–6. IEEE (2017)
14. SENUM: Communication SENUM CAMP, Présidence de la République du Sénégal (2025). https://www.presidence.sn/wo/assets/documents/Communication-SENUM-CAMP-2025.pdf
15. Ndassimba, N.G., Kossingou, G.M., Ndassimba, E., Gueye, K., Ouya, S.: The impact of zeroconf on the education system of the countries in conflict: the case of the central african republic. In: 2020 22nd International Conference on Advanced Communication Technology (ICACT), pp. 614–619. IEEE (2020)
16. Sakib, S.K., Amariucai, G.T., Guan, Y.: Zeroprokes: A secure zeroconf key establishment protocol for large-scale low-cost applications. IEEE Trans. Depend. Sec. Comput. (2022)
17. Holden, C.W., Pierson, M., Wu, J.: In the blink of an eye: exchange-to-sip latency and trade classification accuracy (2023). Available at SSRN 4441422
18. Numerique, S.: Étude sur le partage des infrastructures des télécommunications et des tic au sénégal. Senegal Numerique

Mulatu Numbers as a Product of Three Repdigits in Base h

Kossi Richmond Kakanou, Pagdame Tiebekabe, and Ismaïla Diouf

Abstract The Mulatu numbers, introduced by Mulatu, form a fascinating sequence:

$$4, 1, 5, 6, 11, 17, 28, 45, \ldots$$

$$M_k = M_{k-1} + M_{k-2}.$$

In our study, we focus on identifying the conditions under which Mulatu numbers can be expressed as a product of three repdigits in a base h. Specifically, we examine these conditions for values of within the range, where is a specified upper bound.

Keywords Mulatu numbers · Repdigits · Linear form in logarithms · Baker's method · Reduction method

2020 Mathematics Subject Classification 11B39 · 11J86 · 11Y50 · 11D61

K. R. Kakanou (✉) · P. Tiebekabe · I. Diouf
Laboratoire d'Algébre, de Cryptologie, de Géométrie Algébrique et Applications (LACGAA), Universite Cheikh Anta Diop de Dakar (UCAD), Dakar, Senegal
e-mail: kossirichmond.kakanou@ucad.edu.sn

P. Tiebekabe
e-mail: pagdame.tiebekabe@ucad.edu.sn

I. Diouf
e-mail: ismaila.diouf@ucad.edu.sn

P. Tiebekabe
Faculté des Sciences et Techniques (FST), Département de Mathématiques, Université de Kara, Kara, Togo

© The Author(s), under exclusive license to Springer Nature Switzerland AG 2026
C. T. Gueye et al. (eds.), *Mathematics, Computer Science, Cybersecurity and Artificial Intelligence*, Springer Proceedings in Mathematics & Statistics 536,
https://doi.org/10.1007/978-3-032-17666-0_13

1 Introduction

The *Mulatu numbers*, introduced by Mulatu, form a fascinating sequence defined recursively by

$$M_k = M_{k-1} + M_{k-2},$$

with initial conditions

$$M_0 = 4, \quad M_1 = 1.$$

The first few terms of the sequence are

$$4,\ 1,\ 5,\ 6,\ 11,\ 17,\ 28,\ 45, \ldots$$

A repdigit in a base (where is an integer) is a number represented by a single repeating digit in that base. For example, in base, a repdigit can be expressed as

$$R = d\sum_{i=0}^{n-1} h^i = d\frac{h^n - 1}{h - 1}, \quad 1 \le d \le h - 1,\ n \ge 1,\ h \in \mathbb{Z},\ h \ge 2.$$

Products of repdigits are particularly interesting due to their unique representations and properties.

In our study, we focus on identifying the conditions under which Mulatu numbers can be expressed as a products of three repdigits in a base. Specifically, we examine these conditions for values of within the range, where is a specified upper bound.

For more details on the work related to the determination of the terms of linear recurrent sequences which are repdigits in any base, we refer the reader to the following recent results [1–4].

In this paper, we delve into a fascinating aspect of Mulatu numbers by examining their representation as a product of three repdigits in base h with $h \ge 2$.

The proofs of our main results are based on a double application of Baker's method and on a reduction algorithm using computations based on continued fractions. The method used to determine Mulatu numbers, which are a products of three repdigits is similar to that used by H. Ben Yakkou et al. [1].

The paper is organized as follows: in Sect. 2, we present our main results, Sect. 3 recalls the necessary results for the proofs of our results, and in Sect. 4, we prove our results.

2 Statement of Main Results

In this section, we state all the main results obtained in this paper.

Theorem 1 *Let $h \ge 2$ be an integer. Then the Diophantine equation*

$$M_k = d_1 \frac{h^\ell - 1}{h-1} \cdot d_2 \frac{h^m - 1}{h-1} \cdot d_3 \frac{h^n - 1}{h-1} \tag{1}$$

has only finitely many solutions in integers $k, d_1, d_2, d_3, \ell, m, n$ *such that* $1 \le d_i \le h-1$ *for* $i = 1, 2, 3$ *and* $n \ge m \ge \ell \ge 1$. *Further, we have*

$$n < 1.15 \times 10^{47} (\log h)^9 \quad \textit{and} \quad k < 8.03 \times 10^{47} (\log h)^{10}.$$

Under the notation and assumptions of Theorem 1, if (1) holds for $\big(k, d_1, d_2, d_3, \ell, m, n\big)$, then we write

$$M_k = [a, b, c]_h = a \times b \times c,$$

where

$$a = d_1 \times \frac{h^\ell - 1}{h-1} = \underbrace{\overline{d_1 \cdots d_1}_h}_{\ell-1 \text{ times}}, \quad b = d_2 \times \frac{h^m - 1}{h-1} = \underbrace{\overline{d_2 \cdots d_2}_h}_{m-1 \text{ times}}, \quad \text{and} \quad c = d_3 \times \frac{h^n - 1}{h-1} = \underbrace{\overline{d_3 \cdots d_3}_h}_{n-1 \text{ times}}.$$

In the following theorem, we completely and explicitly give all solutions of the equation (1) corresponding to $2 \le h \le 10$.

Theorem 2 *For* $2 \le h \le 10$, *the only Mulatu numbers that can be written as a product of three repdigits in base h are*

$$\{4, 5, 6, 11, 17, 28, 45, 73, 118, 191, 309, 500, 809, 1309, 2118, 5545\}.$$

More precisely, we have (Table 1).

3 Preliminary Results

In this section, we give some notation and recall certain definitions and results required for the proof of our main results.

3.1 Some Properties of the Mulatu Sequence

The first few terms of the *Mulatu sequence* are as follows:

4, 1, 5, 6, 11, 17, 28, 45, 73, 118, 191, 309, 500, 809, 1309, 2118, ...

This sequence is defined by the recurrence relation

Table 1 Mulatu numbers which are a product of three repdigits in base where $2 \leq h \leq 10$

k	M_k	$[a, b, c]_g$
0	4	$[1, 0, 0]_2$.
2	5	$[1, 0, 1]_2$
3	6	$[1, 1, 0]_2$.
4	11	$[1, 0, 11]_2$, $[1, 0, 2]_3$.
5	17	$[1, 000, 1]_2$, $[1, 2, 2]_3$, $[1, 0, 1]_4$.
6	28	$[1, 00, 1]_3$, $[1, 3, 0]_4$, $[1, 0, 3]_5$.
7	45	$[1, 2, 00]_3$,$[2, 3, 1]_4$,$[1, 4, 0]_5$.
8	73	$[22, 0, 1]_3$, $[2, 4, 3]_5$, $[2, 0, 1]_6$.
9	118	$[111, 0, 1]_3$, $[3, 1, 4]_6$, $[1, 4, 1]_9$.
10	191	$[1, 0, 111111]_2$, $[5, 1, 5]_6$, $[3, 6, 2]_7$, $[1, 9, 2]_9$, $[1, 9, 12]_{10}$.
11	309	$[22, 1, 4]_5$, $[1, 2, 33]_6$, $[6, 2, 1]_7$, $[4, 6, 5]_8$, $[3, 7, 3]_9$, $[3, 0, 9]_{10}$.
12	500	$[7, 6, 4]_8$, $[6, 1, 5]_9$.
13	809	$[1, 00, 2222]_3$, $[22, 3, 4]_7$, $[1, 0, 88]_9$, $[8, 0, 9]_{10}$.
14	1309	$[3, 55, 0]_7$.
15	2118	$[6, 11, 4]_7$, $[2, 11, 8]_{10}$.
17	5545	$[111, 222, 1]_4$, $[1, 3, 333]_4$, $[55, 4, 5]_{10}$.

$$M_k = M_{k-1} + M_{k-2},$$

with initial conditions

$$M_0 = 4, \quad M_1 = 1.$$

Its characteristic equation is given by

$$x^2 - x - 1 = 0.$$

The roots of this equation are given by

$$\alpha = \frac{1 + \sqrt{5}}{2} \quad \text{and} \quad \beta = \frac{1 - \sqrt{5}}{2}.$$

Using these roots, Binet's formula for M_k can be expressed as

$$M_k = a\alpha^k + b\beta^k, \quad \text{for } k \geq 0, \tag{2}$$

where $a = (\frac{10 - \sqrt{5}}{5})$ and $b = (\frac{10 + \sqrt{5}}{5})$.

Furthermore, we have

$$\alpha^{k-1} \leq M_k \leq 4\alpha^k, \quad \text{for } k \geq 0. \tag{3}$$

3.2 Linear Forms in Logarithms

We begin this subsection with a few reminders about the logarithmic height of an algebraic number. Let η be an algebraic number of degree d, $a_0 > 0$ be the leading coefficient of its minimal polynomial over $\mathbb{Z}$, and let $\eta = \eta^{(1)}, \ldots, \eta^{(d)}$ denote its conjugates. The quantity defined by

$$LH(\eta) = \frac{1}{d}\left(\log|a_0| + \sum_{j=1}^{d} \log\max(1, |\eta^{(j)}|)\right)$$

is called the logarithmic height of η. Some properties of height are as follows. For η_1, η_2 algebraic numbers and $m \in \mathbb{Z}$, we have

$$\begin{aligned} LH(\eta_1 \pm \eta_2) &\leq LH(\eta_1) + LH(\eta_2) + \log 2, \\ LH(\eta_1 \cdot \eta_2^{\epsilon}) &\leq LH(\eta_1) + LH(\eta_2), \text{ where } \epsilon = \pm 1 \\ LH(\eta_1^m) &= |m| LH(\eta_1). \end{aligned}$$

In particular, if $\eta = p/q \in \mathbb{Q}$ is a rational number in its reduced form with $q > 0$, then $LH(\eta) = \log(\max\{|p|, q\})$.

We can now present the well-known result of Matveev used in this study. Let $\mathbb{L}$ be a real number field of degree $d_{\mathbb{L}}$, $\eta_1, \ldots, \eta_s \in \mathbb{L}$, and $b_1, \ldots, b_s \in \mathbb{Z} \setminus \{0\}$. Let $B \geq \max\{|b_1|, \ldots, |b_s|\}$ and

$$\Lambda = \eta_1^{b_1} \cdots \eta_s^{b_s} - 1.$$

Let $A_1, \ldots, A_s$ be real numbers such that

$$A_i \geq \max\{d_{\mathbb{L}} LH(\eta_i), |\log \eta_i|, 0.16\}, \quad i = 1, \ldots, s.$$

With the above notation, Matveev [6] proved the following result.

Theorem 3 *Assume that $\Lambda \neq 0$. Then*

$$\log|\Lambda| > -1.4 \cdot 30^{s+3} \cdot s^{4.5} \cdot d_{\mathbb{L}}^2 \cdot (1 + \log d_{\mathbb{L}}) \cdot (1 + \log B) \cdot A_1 \cdots A_s.$$

We also need the following result of Sanchez and Luca [7].

Lemma 1 *Let $r \geq 1$ and $H > 0$ be such that $H > (4r^2)^r$ and $H > L/(\log L)^r$. Then*

$$L < 2^r H (\log H)^r.$$

3.3 Reduction Method

The bounds on the variables obtained via Baker's theory [8] are too large for any computational purposes. To reduce the bounds, we use the reduction method due to Dujella and Pethő [9, Lemma 5a]. For a real number X, we set $\|X\| := \min\{|X - n| : n \in \mathbb{Z}\}$ to denote the distance from X to the nearest integer.

Lemma 2 *Let M be a positive integer, p/q be a convergent of the continued fraction expansion of an irrational number τ such that $q > 6M$, and A, B, μ be some real numbers with $A > 0$ and $B > 1$. Furthermore, set*

$$\varepsilon := \|\mu q\| - M \cdot \|\tau q\|.$$

If $\varepsilon > 0$, then there is no solution to the inequality

$$0 < |u\tau - v + \mu| < AB^{-w} \tag{4}$$

in positive integers u, v and w with

$$u \leq M \text{ and } w \geq \frac{\log(Aq/\varepsilon)}{\log B}.$$

4 Proofs of Main Results

4.1 Proof of Theorem 1

To prove Theorem 1, we will use the following lemma which provides a relation on the size of k versus n and g.

Lemma 3 *All solutions of the Diophantine equation* (1) *satisfy*

$$k < 7n \log h.$$

Proof From the inequality

$$\alpha^{k-1} \leq M_k \leq (h^n - 1)^3 < h^{3n},$$

we obtain

$$(k-1)\log\alpha < 3n\log h \quad \Longrightarrow \quad k < 1 + \frac{3n\log h}{\log\alpha}.$$

Numerically, we have $\log\alpha = 0.481211825\ldots$, hence

$$\frac{3}{\log\alpha} = 6.23426076\ldots,$$

and therefore

$$k < 1 + 6.23426\, n \log h.$$

□

Proof of Theorem 1 If $n = 1$, then $\ell = m = 1$. So equation (1) becomes

$$M_k = d_1 d_2 d_3,$$

which implies

$$\alpha^{k-1} \leq (h-1)^3,$$

which leads to

$$k < 1 + 3\frac{\log h}{\log \alpha}.$$

Now suppose $n \geq 2$. From (1) and (2), we have

$$M_k = a\alpha^k + b\beta^k = d_1 \frac{h^\ell - 1}{h-1} \cdot d_2 \frac{h^m - 1}{h-1} \cdot d_3 \frac{h^n - 1}{h-1},$$

which implies

$$\begin{aligned} a\alpha^k - \frac{d_1 d_2 d_3 h^{\ell+m+n}}{(h-1)^3} &= -\frac{d_1 d_2 d_3 \left(h^{\ell+m} + h^{\ell+n} + h^{m+n}\right)}{(h-1)^3} \\ &\quad + \frac{d_1 d_2 d_3 \left(h^{l} + h^m + h^n\right)}{(h-1)^3} - \frac{d_1 d_2 d_3}{(h-1)^3} - b\beta^k. \end{aligned} \tag{5}$$

Taking the absolute values of both sides of (5), we get

$$\begin{aligned} \left| a\alpha^k - \frac{d_1 d_2 d_3 h^{\ell+m+n}}{(h-1)^3} \right| &< \frac{d_1 d_2 d_3 \left(h^{\ell+m} + h^{\ell+n} + h^{m+n}\right)}{(h-1)^3} \\ &\quad + \frac{d_1 d_2 d_3 \left(h^{l} + h^m + h^n\right)}{(h-1)^3} + \frac{d_1 d_2 d_3}{(h-1)^3} + \frac{b}{\alpha^k}. \end{aligned} \tag{6}$$

Multiplying both sides of (6) by $\dfrac{(h-1)^3}{d_1 d_2 d_3 h^{\ell+n+m}}$ and noticing the fact that $\ell \leq m \leq n$, we get the inequality

$$\left|\frac{(h-1)^3 \cdot a\alpha^k \cdot h^{-(\ell+n+m)}}{d_1d_2d_3} - 1\right| < \frac{1}{h^\ell} + \frac{1}{h^m} + \frac{1}{h^n} + \frac{1}{h^{\ell+m}} + \frac{1}{h^{\ell+n}} + \frac{1}{h^{\ell+m+n}} + \frac{b(h-1)^3}{\alpha^k d_1d_2d_3h^{\ell+n+m}} < 7 \cdot h^{-\ell}.$$

So we get

$$\left|\frac{a(h-1)^3}{d_1d_2d_3} \cdot \alpha^k \cdot h^{-(\ell+n+m)} - 1\right| < 7 \cdot h^{-\ell}. \tag{7}$$

We set

$$\Gamma_1 = \frac{a(h-1)^3}{d_1d_2d_3} \cdot \alpha^k \cdot h^{-(\ell+n+m)} - 1.$$

Let us show $\Gamma_1 \neq 0$. We proceed by the contradiction. Assume that $\Gamma_1 = 0$. Then

$$a\alpha^k = \frac{d_1d_2d_3}{(h-1)^3} \cdot h^{\ell+m+n},$$

which implies

$$\sigma_\varphi\left(a\alpha^k\right) = b\beta^k = \frac{d_1d_2d_3}{(h-1)^3} \cdot h^{\ell+m+n},$$

where σ_φ is the permutation $(\alpha\beta)$.

Taking the absolute values, we obtain

$$\left|b\beta^k\right| = \left|\frac{d_1d_2d_3}{(h-1)^3} \cdot h^{\ell+m+n}\right|.$$

We have $\left|b\beta^k\right| < 1$ instead $\left|\frac{d_1d_2d_3}{(h-1)^3} \cdot h^{\ell+m+n}\right| > 1$ since $1 \leq \ell \leq m \leq n$, which leads to a contradiction. Hence $\Gamma_1 \neq 0$.

In order to apply Matveev's result to Γ_1, set

$$s := 3, \quad \eta_1 := \frac{a(h-1)^3}{d_1d_2d_3}, \quad \eta_2 := \alpha, \quad \eta_3 := h,$$
$$b_1 := 1, \quad b_2 := k, \quad b_3 := -(\ell+m+n),$$

and $\mathbb{K} := \mathbb{Q}(\eta_1, \eta_2, \eta_3) = \mathbb{Q}(\alpha)$ which is a real number field of degree $d_{\mathbb{K}} = 2$.

Using properties of the logarithmic height, we get

$$LH(\eta_2) = LH(\alpha) = \frac{\log \alpha}{2}, \quad LH(\eta_3) = LH(h) = \log h$$

and

$$\begin{aligned} LH(\eta_1) &= LH\left(\frac{a(h-1)^3}{d_1 d_2 d_3}\right) \\ &\leq LH(a) + LH\left(\frac{(h-1)^3}{d_1 d_2 d_3}\right) \\ &\leq \frac{1}{2}\log 19 + \log\left(\max\left\{(h-1)^3, d_1 d_2 d_3\right\}\right) \\ &< 3\log h + 2 \\ &< 6\log h, \end{aligned}$$

since $h \geq 2$. Thus, we can take

$$A_1 = 12\log h, \quad A_2 = \log \alpha, \quad \text{and } A_3 = 2\log h.$$

By Lemma 3, we have $k < 7n\log h$, so we put $B = 7n\log h$. Using Theorem 3, we see that

$$\begin{aligned} \log|\Gamma_1| >& -1.4 \times 30^6 \times 3^{4.5} \times 2^2(1+\log 2)(1+\log(7n\log h)) \\ &\times (12\log h)(2\log h\log\alpha) \\ >& -1.12 \times 10^{13}(1+\log(7n\log h))(\log h)^2. \end{aligned}$$

Comparing the above inequality with (7), we obtain that

$$\ell\log h - \log 7 < 1.12 \times 10^{13}(1+\log(7n\log h))(\log h)^2.$$

Since $h \geq 2$ and $n \geq 2$, we have

$$1 + \log(7n\log h) < 7n\log h,$$

so we get

$$\ell < 7.85 \times 10^{13} n \log(h)^2.$$

Rewriting (1), we get

$$\frac{a\alpha^k(h-1)}{d_1(h^\ell - 1)} + \frac{b\beta^k(h-1)}{d_1(h^\ell - 1)} = \frac{d_2 d_3}{(h-1)^2}\left(h^{n+m} - h^m - h^n + 1\right),$$

which implies

$$\frac{a\alpha^k(h-1)}{d_1(h^\ell-1)} - \frac{d_2d_3h^{n+m}}{(h-1)^2} = -\frac{b\beta^k(h-1)}{d_1(h^\ell-1)} - \frac{d_2d_3h^m}{(h-1)^2} - \frac{d_2d_3h^n}{(h-1)^2} + \frac{d_2d_3}{(h-1)^2}. \tag{8}$$

Taking the absolute values of both sides of (8), we have

$$\left|\frac{a\alpha^k(h-1)}{d_1(h^\ell-1)} - \frac{d_2d_3h^{n+m}}{(h-1)^2}\right| < \frac{b(h-1)}{d_1(h^\ell-1)\alpha^k} + \frac{d_2d_3h^m}{(h-1)^2} + \frac{d_2d_3h^n}{(h-1)^2} + \frac{d_2d_3}{(h-1)^2}.$$

Dividing both sides of the inequality above by $\frac{d_2d_3h^{n+m}}{(h-1)^2}$ and using the fact that $n \geq 2$, we see that

$$\begin{aligned}\left|\frac{(h-1)^3}{d_1d_2d_3(h^\ell-1)} \cdot a\alpha^k \cdot h^{-(n+m)} - 1\right| &\leq \frac{b(h-1)^3}{d_1d_2d_3(h^\ell-1)\alpha^k h^{n+m}} + \frac{1}{h^n} + \frac{1}{h^m} + \frac{1}{h^{n+m}} \\ &< 4 \cdot h^{-m}.\end{aligned}$$

Then we have

$$\left|\frac{(h-1)^3}{d_1d_2d_3(h^\ell-1)} \cdot a\alpha^k \cdot h^{-(n+m)} - 1\right| < \frac{4}{h^m}. \tag{9}$$

We put

$$\Gamma_2 = \frac{(h-1)^3}{d_1d_2d_3(h^\ell-1)} \cdot a\alpha^k \cdot h^{-(n+m)} - 1.$$

One can check that $\Gamma_2 \neq 0$, proceeding as we do for Γ_1. Let us apply Matveev's result for Γ_2. Set

$$\begin{aligned}&s := 3, \quad \eta_1 := \frac{(h-1)^3}{d_1d_2d_3(h^\ell-1)} \cdot a, \quad \eta_2 := \alpha, \quad \eta_3 := g, \\ &b_1 := 1, \quad b_2 := k, \quad b_3 := -(m+n),\end{aligned}$$

and $\mathbb{K} := \mathbb{Q}(\eta_1, \eta_2, \eta_3) = \mathbb{Q}(\alpha)$ of degree $d_{\mathbb{K}} = 3$. By Lemma 3, we have $k < 7n \log h$, so we put $B = 7n \log h$. We have

$$LH(\eta_2) = LH(\alpha) = \frac{\log \alpha}{2}, \quad LH(\eta_3) = LH(h) = \log h,$$

and

$$\begin{aligned}
LH(\eta_1) &= h\left(\frac{(h-1)^3}{d_1d_2d_3(h^\ell-1)}\cdot a\right)\\
&\le LH(a) + LH\left(\frac{(h-1)^3}{d_1d_2d_3(h^\ell-1)}\right)\\
&\le \frac{1}{2}\log 19 + \log\left(\max\left\{(h-1)^3, d_1d_2d_3\right\}\right) + LH\left(\frac{1}{h^\ell-1}\right)\\
&< 2 + 3\log(h-1) + \log(h^\ell-1)\\
&< (3+\ell)\log h + 2\\
&< (6+\ell)\log h \quad \text{since } h \ge 2.
\end{aligned}$$

Thus, we can take

$$A_1 = (12+2\ell)\log h, \quad A_2 = \log\alpha, \quad \text{and } A_3 = 2\log h.$$

Using Theorem 3, we see that

$$\begin{aligned}
\log|\Gamma_2| > &-1.4\times 30^6\times 3^{4.5}\times 2^2(1+\log 2)(1+\log(7n\log h))\\
&\times((12+2\ell)\log h)(2\log h\log\alpha)\\
&> -4.27\times 10^{11}(1+\log(7n\log h))(\log h)^2(12+2\ell).
\end{aligned}$$

Comparing with (9), we get

$$m\log h - \log 4 < 4.27\times 10^{11}(1+\log(7n\log h))(\log h)^2(12+2\ell).$$

We have

$$1+\log(7n\log h) < 7n\log h \quad \text{and} \quad \ell < 7.85\times 10^{13} n(\log h)^2.$$

It follows that

$$m < 4.7\times 10^{26}(n)^2(\log h)^4.$$

Reorganizing (1), we get

$$\frac{d_3h^n}{h-1} - \frac{(h-1)^2\cdot a\alpha^k}{d_1d_2(h^\ell-1)(h^m-1)} = \frac{d_3}{h-1} - \frac{b\beta^k(h-1)^2}{d_1d_2(h^\ell-1)(h^m-1)}.$$

We have

$$\left|\frac{d_3h^n}{h-1} - \frac{(h-1)^2\cdot a\alpha^k}{d_1d_2(h^\ell-1)(h^m-1)}\right| < \frac{d_3}{h-1} + \frac{b(h-1)^2}{\alpha^k d_1d_2(h^\ell-1)(h^m-1)}$$

by taking the absolute values of both sides of (8). Dividing both sides of the above inequality by $(d_3h^n)/(h-1)$ and using the fact that $n \geq 2$, we see that

$$\left|1 - \frac{a(h-1)^3}{d_1d_2d_3(h^\ell - 1)(h^m - 1)} \cdot h^{-n} \cdot \alpha^k\right| < \frac{1}{h^n} + \frac{1}{h^{n-1}} < \frac{2}{h^{n-1}}.$$
$$< 2 \cdot h^{1-n}.$$

Then we have

$$\left|\frac{a(h-1)^3}{d_1d_2d_3(h^\ell - 1)(h^m - 1)} \cdot h^{-n} \cdot \alpha^k - 1\right| < 2 \cdot h^{1-n}. \tag{10}$$

We set

$$\Gamma_3 := \frac{a(h-1)^3}{d_1d_2d_3(h^\ell - 1)(h^m - 1)} \cdot h^{-n} \cdot \alpha^k - 1.$$

One can verify that $\Gamma_3 \neq 0$. Let us analyze Matveev's result for Γ_3. Set

$$s := 3, \quad \eta_1 := \frac{a(h-1)^3}{d_1d_2d_3(h^\ell - 1)(h^m - 1)}, \quad \eta_2 := \alpha, \quad \eta_3 := g,$$
$$b_1 := 1, \quad b_2 := k, \quad b_3 := -n,$$

and $\mathbb{K} := \mathbb{Q}(\eta_1, \eta_2, \eta_3) = \mathbb{Q}(\alpha)$ of degree $d_{\mathbb{K}} = 2$. By Lemma 3, we have $k < 7n \log h$, so we put $B = 7n \log h$. We have

$$LH(\eta_2) = LH(\alpha) = \frac{\log \alpha}{2}, \quad LH(\eta_3) = LH(h) = \log h,$$

and

$$\begin{aligned}
LH(\eta_1) &= LH\left(\frac{a(h-1)^3}{d_1d_2d_3(h^\ell - 1)(h^m - 1)}\right) \\
&\leq LH(a) + LH\left(\frac{(h-1)^3}{d_1d_2d_3(h^\ell - 1)(h^m - 1)}\right) \\
&\leq \frac{1}{2}\log 19 + \log\left(\max\{(h-1)^3, d_1d_2d_3\}\right) + LH\left(\frac{1}{h^\ell - 1}\right) + LH\left(\frac{1}{h^m - 1}\right) \\
&< 2 + 3\log\left(h-1\right) + \log\left(h^\ell - 1\right) + \log\left(h^m - 1\right) \\
&< (3 + \ell + m)\log h + 2 \\
&< (6 + \ell + m)\log h,
\end{aligned}$$

since $h \geq 2$, we can take

$$A_1 = 2(6 + \ell + m)\log h, \quad A_2 = \log \alpha, \quad \text{and } A_3 = 2\log h.$$

Using Theorem 3, we see that

$$\begin{aligned}\log\left|\Gamma_3\right| &> -1.4 \times 30^6 \times 3^{4.5} \times 2^4(1+\log 2)(1+\log(7n\log h)) \\ &\quad \times ((6+\ell+m)\log h)(\log h \log \alpha) \\ &> -1.94 \times 10^{12}(1+\log(7n\log h))(\log h)^2(6+\ell+m).\end{aligned}$$

Comparing with (10), we get

$$(n-1)\log h - \log 2 < 1.94 \times 10^{12}(1+\log(7n\log h))(\log h)^2(6+\ell+m).$$

We have

$$\begin{aligned}&1+\log(7n\log h) < 7n\log h, \quad m < 4.7 \times 10^{26} \times (n)^2(\log h)^4 \\ &\text{and} \quad \ell < 7.85 \times 10^{13} n(\log h)^2.\end{aligned}$$

It follows that

$$\begin{aligned}6+\ell+m &< 7.85 \times 10^{13} n(\log h)^2 + 4.7 \times 10^{26} \times (n)^2(\log h)^4 \\ &< 4.8 \times 10^{26} \times (n)^2(\log h)^4.\end{aligned}$$

So we have

$$n < 6.52 \times 10^{39}(n)^3(\log h)^6.$$

Now we apply Lemma 1, by setting

$$r := 3, \quad L := n \quad \text{and} \quad H := 3 \times 10^{42} \cdot (\log h)^6,$$

we get

$$\begin{aligned}n &< 2^3 \cdot 6.52 \times 10^{39} \cdot (\log h)^6 \times \left(\log\left(6.52 \times 10^{39} \cdot (\log h)^6\right)\right)^3 \\ &< 5.216 \times 10^{40} \cdot (\log h)^6 \cdot (91.68 + 6\log\log h)^3 \\ &< 1.15 \times 10^{47}(\log h)^9.\end{aligned}$$

Notice that we have used the inequality $95.6 + 6\log\log h < 130\log h$ which holds since $h \geq 2$. □

4.2 Proof of Theorem 2

Since $2 \le h \le 10$, according to Theorem 1, we have

$$\ell \le m \le n < 2.09 \times 10^{50} \quad \text{and} \quad k < 3.37 \times 10^{51}.$$

Consequently, the next step is to reduce the upper bounds above in order to identify the set of the interval in which the possible solutions of (1) lie. To do this, we proceed in three steps.

Step 1 Using (7), set

$$\Lambda_1 := -\log(\Gamma_1 + 1) = (\ell + m + n)\log h - k\log\alpha - \log\left(\frac{(h-1)^3 a}{d_1 d_2 d_3}\right).$$

Notice that (7) can be rewritten as

$$\left|e^{-\Lambda_1} - 1\right| < \frac{7}{h^{\ell}}.$$

Observe that $\Lambda_1 \neq 0$, since $e^{-\Lambda_1} - 1 = \Gamma_1 \neq 0$. Assume that $\ell \ge 5$. Then

$$\left|e^{-\Lambda_1} - 1\right| < \frac{7}{h^{\ell}} < \frac{1}{2}.$$

Since $|x| < 2|e^x - 1|$, if $|x| < \frac{1}{2}$ holds, then

$$\left|\Lambda_1\right| < \frac{14}{h^{\ell}}.$$

Substituting Λ_1 in the above inequality with its value and dividing through by $\log\alpha$, we get

$$\left|(\ell + m + n)\left(\frac{\log h}{\log\alpha}\right) - k + \frac{\log\left(\frac{(h-1)^3 a}{d_1 d_2 d_3}\right)}{\log\alpha}\right| < \frac{14}{\log\alpha \cdot h^{\ell}}.$$

Then, we can apply Lemma 2 with the data

$$\tau := \frac{\log h}{\log\alpha}, \quad \mu := \frac{\log\left(\frac{(h-1)^3 a}{d_1 d_2 d_3}\right)}{\log\alpha}, \quad A := \frac{14}{\log\alpha},$$
$$B := h, \quad \text{and} \quad w := \ell \quad \text{with} \quad 1 \le d_1 \le d_2 \le d_3 \le h - 1.$$

Table 2 Upper bound on ℓ

g	2	3	4	5	6	7	8	9	10
q_t	q_{108}	q_{94}	q_{106}	q_{108}	q_{106}	q_{103}	q_{100}	q_{86}	q_{112}
$\varepsilon \geq$	0.34	0.37	0.18	0.05	0.48	0.35	0.16	0.42	0.34
$\ell \leq$	183	114	92	79	72	66	61	57	55

We can take $M := 3.37 \times 10^{51}$, since $k < 7n \log h < 3.37 \times 10^{51}$. So, for the remaining proof, we use *Mathematica* to apply Lemma 2. For the computations, if the first convergent q_t is such that $q_t > 6M$ does not satisfy the condition $\varepsilon > 0$, then we use the next convergent until we find the one that satisfies the conditions. Thus, we have that (Table 2).

It follows that

$$1 \leq \ell \leq \frac{\log((14/\log\alpha) \cdot q_{108}/0.34)}{\log 2} \leq 183.$$

Step 2 In this step, we have to reduce the upper bound on m. To do this, let us consider

$$\Lambda_2 := -\log(\Gamma_2 + 1) = (m+n)\log h - k\log\alpha + \log\left(\frac{(h-1)^3 a}{d_1 d_2 d_3 \cdot (h^\ell - 1)}\right).$$

Thus, inequality (9) becomes

$$\left|e^{-\Lambda_2} - 1\right| < \frac{4}{h^m} < \frac{1}{2},$$

which holds for $m \geq 4$. It follows that

$$\left|(m+n)\frac{\log h}{\log\alpha} - k + \frac{\log\left(\frac{(h-1)^3 a}{d_1 d_2 d_3 \cdot (h^\ell - 1)}\right)}{\log\alpha}\right| < \frac{7}{h^m \log\alpha}. \tag{11}$$

So, the conditions of Lemma 2 are satisfied. Applying this lemma to Inequality (11) with the following data

$$\tau := \frac{\log h}{\log\alpha}, \quad \mu := \frac{\log\left(\frac{(h-1)^3 a}{d_1 d_2 d_3 \cdot (h^\ell - 1)}\right)}{\log\alpha}, \quad A := \frac{7}{\log\alpha}, \quad B := h, \quad \text{and} \quad w := m$$

with $1 \leq d_1 \leq d_2 \leq d_3 \leq h-1$ and $1 \leq \ell \leq 183$.

As $k < 7n \log h < 3.37 \times 10^{51}$, we can take $M := 3.37 \times 10^{51}$. With *Mathematica* we get the following results (Table 3):

Table 3 Upper bound on m

g	2	3	4	5	6	7	8	9	10
q_t	q_{108}	q_{95}	q_{106}	q_{106}	q_{106}	q_{103}	q_{101}	q_{86}	q_{112}
$\varepsilon \geq$	0.3	0.04	0.15	0.014	0.01	0.005	0.001	0.003	0.004
$m \leq$	189	118	94	81	74	66	63	59	57

In all cases, we can conclude that

$$1 \leq m \leq \frac{\log((7/\log\alpha)\cdot q_{108}/0.3)}{\log 2} \leq 189.$$

Step 3 Finally, to reduce the bound on n we have to choose

$$\Lambda_3 := \log(\Gamma_3 + 1) = (n)\log h - k\log\alpha + \log\left(\frac{(h-1)^3 a}{d_1d_2d_3\cdot(h^\ell - 1)(h^m - 1)}\right).$$

We have

$$\left|e^{-\Lambda_3} - 1\right| < \frac{2}{h^{n-1}} < \frac{1}{2},$$

which is valid for $n \geq 4$ and $g \geq 2$. It follows that

$$\left|m\frac{\log h}{\log\alpha} - k + \frac{\log\left(\frac{(h-1)^3 a}{d_1d_2d_3\cdot(h^\ell - 1)(h^m - 1)}\right)}{\log\alpha}\right| < \frac{4}{h^{n-1}\alpha}. \tag{12}$$

Now we have to apply Lemma 2 to (12) by taking the following parameters:

$$\tau := \frac{\log h}{\log\alpha}, \quad \mu := \frac{\log\left(\frac{(h-1)^3 a}{d_1d_2d_3\cdot(h^\ell - 1)(h^m - 1)}\right)}{\log\alpha}, \quad A := \frac{7}{\log\alpha},$$
$$B := g, \quad \text{and} \quad w := n - 1$$

with $1 \leq d_1 \leq d_2 \leq d_3 \leq h - 1$, $1 \leq \ell \leq 189$, and $1 \leq m \leq 183$.

Using the fact that $k < 7n\log h < 1.99 \times 10^{54}$, we can take $M := 3.37 \times 10^{51}$, and we get results in Table 4.

It follows from Table 4 that

$$1 \leq n \leq \frac{\log((4/\log\alpha)\cdot q_{118}/0.000002)}{\log 2} \leq 192,$$

Table 4 Upper bound on n

g	2	3	4	5	6	7	8	9	10
q_t	q_{108}	q_{95}	q_{107}	q_{109}	q_{106}	q_{104}	q_{100}	q_{102}	q_{112}
$\varepsilon \geq$	0.3	0.04	0.04	0.012	0.004	0.000002	0.002	0.02	0.003
$n \leq$	192	120	95	81	74	72	63	61	57

which is valid for all h such that $2 \leq h \leq 10$. In light of the above results, we need to check the equation (1) in the cases $2 \leq h \leq 10$ for $1 \leq d_1, d_2, d_3 \leq 9$, $1 \leq n \leq 192$, $1 \leq m \leq 189$, $1 \leq \ell \leq 183$ and $1 \leq k \leq 7460$. A quick inspection using *SageMath* reveals that the Diophantine equation (1) in the range $2 \leq h \leq 10$ has only the solution listed in the statement of Theorem 2. This completes the proof of Theorem 2.

Acknowledgements The authors express their gratitude to the anonymous reviewer for the instructive suggestions. The first and the third authors are supported by UCAD, the Université Cheikh Anta Diop de Dakar, and the second author is partially supported by Université de Kara (Togo). This project was initiated when the second author visited UCAD on a research stay. He thanks the authorities for the warm hospitality and the working environment.

References

1. Ben Yakkou, H., Kakanou, K.R., Tiebekabe, P.: Narayana numbers as products of three repdigits in base g. Acta et Commentationes Universitatis Tartuensis de Mathematica **27**(2), (2023). https://ojs.utlib.ee/index.php/ACUTM
2. Adédji,K.N.: Balancing numbers which are products of three repdigits in base b. Bol. Soc. Mat. Mex. **29** (2023)
3. Adédji, K.N., Filipin, A., Togbé, A.: Fibonacci and Lucas numbers as products of three repdigits in base g. Rend. Circ. Mat. Palermo, II. Ser. (2023)
4. Tiebekabe, P., Adédji, K.N.: Narayana's cows numbers which are concatenations of three repdigits in base ρ (2023). arXiv:2304.00773
5. Allouche, J.P., Johnson, T.: Narayana's cows and delayed morphisms, J. d'Inform. Music. (1996). HAL:hal-02986050
6. Matveev, E.M.: An explicit lower bound for a homogeneous rational linear form in logarithms of algebraic numbers II. Izv. Math. **64**, 1217–1269 (2000)
7. Sanchez, S.G., Luca, F.: Linear combinations of factorials and S-units in a binary recurrence sequence. Ann. Math. Québec. **38**, 169–188 (2014)
8. Baker, A., Davenport, H.: The equations $3x^2 - 2 = y^2$ and $8x^2 - 7 = z^2$. Q. J. Math. **20**, 129–137 (1969)
9. Dujella, A., Pethő, A.: A generalization of a theorem of Baker and Davenport. Q. J. Math. **49**, 291–306 (1998)
10. Bugeaud, Y., Mignotte, M., Siksek, S. : Classical and modular approaches to exponential Diophantine equations I. Fibonacci and Lucas perfect powers. Ann. Math. **163**, 969–1018 (2006)
11. Irmak, N.: On factorials in Perrin and Padovan sequences. Turk. J. Math. **43**, 2602–2609 (2019)
12. Schlickewei, H.P., Schmidt, W.M.: Linear equations in members of recurrence sequences. Ann. Sc. Norm. Super. Pisa, Cl. Sci., IV. Ser., **20**, 219–246 (1993)

13. Sloane, N.J.A. et al.: The On-Line Encyclopedia of Integer Sequences (2023)

MIX
Papier aus verantwortungsvollen Quellen
Paper from responsible sources
FSC® C105338

If you have any concerns about our products,
you can contact us on
ProductSafety@springernature.com

In case Publisher is established outside the EU,
the EU authorized representative is:
Springer Nature Customer Service Center GmbH
Europaplatz 3, 69115 Heidelberg, Germany

Printed by Libri Plureos GmbH
in Hamburg, Germany